Astronomers
AT HERSTMONCEUX

in their own words

GW00543662

Astronomers
AT HERSTMONCEUX

in their own words

Peter Corben
Dorothy Hobden
Derek Jones
Bill Nicholson
Brian Scales
Rosemary Selmes
Roy Wallis
George Wilkins

Anthony Wilson (editor)

Science Projects Publishing

© The Contributors and Editor 1999
Reprinted 2006, 2017

All rights reserved. No part of this publication may be reproduced, stored in a retrieval system, or transmitted in any form or by any means, electronic, mechanical, photocopying, recording or otherwise, without the prior permission of the publisher.

British Library Cataloguing-in-Publication Data
A catalogue record for this book is available from the British Library.

ISBN 0 9512394 1 4

Published by Science Projects Publishing and obtainable from:
The Observatory Science Centre
Wartling Road, Herstmonceux, East Sussex BN27 1RN
T 01323 832731 **E** info@the-observatory.org
www.the-observatory.org

Cover photograph: The 36-inch Yapp reflector in Dome B of the Equatorial Group, photographed in 1974, operated by Paul Murdin. (RGO photograph)

Contents

Illustrations

The Contributors

Peter Corben spent 32 years at Herstmonceux, after four years at the Royal Observatory, Greenwich. He worked on the 26-inch and 13-inch refractors, the 36-inch reflector and the Isaac Newton Telescope. During this time he also spent several years at the other two Royal Observatories, in Capetown and Edinburgh, and at Siding Springs, Australia and La Palma.

Dorothy Hobden worked for 29 years in Her Majesty's Nautical Almanac Office, the GALAXY Department (automated plate-measuring system), and the La Palma Computing Department, mainly in programming and systems work. She took early retirement when the Observatory moved to Cambridge in 1990.

Derek Jones observed in the Equatorial Group between 1958 and 1988, with intervals away in the USA, South Africa, Australia and the Canary Islands. He was on the project team for the British Observatory in the Canary Islands from its beginning and was also a Visiting Lecturer at the University of Sussex.

Bill Nicholson worked at Herstmonceux for 32 years, joining Her Majesty's Nautical Almanac Office in 1954, and transferring to the Astrometry Department in 1966. He retired in 1986.

Brian Scales transferred from the Met. Office (Air Ministry) to Herstmonceux in 1956 and worked in the Meridian and Astrometry Departments. He was part of the first group to go to the new observatory in La Palma in 1982, and retired at the end of his tour of duty there in 1985.

Rosemary Selmes spent seven years working on the Quasar Research Programme at Herstmonceux. This involved working as a Night Observer on the Thompson 26-inch refractor.

Roy Wallis worked at the RGO from 1954 to 1991 in the Time, Astrophysics and Astrometry departments, and in the Galactic & Extra-galactic Research Group. This included tours of duty in South Africa and La Palma (where he doubled as Publicity Officer).

Dr George Wilkins joined the Observatory in 1951, immediately after leaving Imperial College, and stayed until he retired in 1989. He became Head of the Almanacs and Time Division in 1974.

Anthony Wilson (editor) was formerly Head of Education and Publishing Manager at the Science Museum in London, and later Education Adviser to The Observatory Science Centre at Herstmonceux.

Preface

FOR FORTY YEARS, Herstmonceux in Sussex was the home of one of the world's leading astronomical establishments, the Royal Greenwich Observatory. Its two most striking features were the splendid mediaeval castle, and the six green domes of a unique set of buildings known as the 'Equatorial Group'. At its height, more than 200 people worked at the observatory and lived in the local community.

In 1990, for various reasons, the observatory moved on, to new accommodation in Cambridge. The new establishment in Cambridge was quite different from the one in Sussex. A purpose-built office block replaced the ancient castle, and no observing took place there. Astronomy itself had changed, and much of what had been done at Herstmonceux was done elsewhere or not at all. The technicians and astronomers of the Royal Greenwich Observatory in Cambridge served and used telescopes in other parts of the world. The sort of institution that Herstmonceux was, and the way of life of the people who worked there, had gone for ever.

For some years the Herstmonceux site was left almost deserted – an astronomical *Marie Celeste* marooned in the Sussex countryside. Eventually, however, new uses were found for it. The castle, with other buildings, became the International Study Centre of Queen's University, Ontario, Canada, while the Equatorial Group opened to the public as Herstmonceux Science Centre, later renamed The Observatory Science Centre.

When the Science Centre began operations in 1995 two things soon became apparent. The first was that many visitors, while they enjoyed the new 'hands-on experience' provided there, were also fascinated by the site itself. What was a deserted observatory, with its domes and telescopes still in place, doing in rural Sussex? Like visitors to Stonehenge they wondered what the observatory was for, and what sort of people built and operated it.

Secondly it was apparent that the answers to these questions could be found, most vividly, in the memories of people who had once worked in the observatory. Several of these people now help in the running of the Science Centre, and others come back to visit us from time to time. The stories they have to tell are always interesting and illuminating.

In 1997 I invited a small number of people who had worked at Herstmonceux – too many would have led to unnecessary duplication – to set down something of their personal experiences, trying in particular to capture the unique flavour of the place. Eight people were good enough to respond, and their varied contributions form the basis of this book. (To others who might have wished to add something, I can only apologise and invite them to send contributions to me at the Science Centre. It may be possible to include these contributions in a second edition of the book; if not, they will be offered to the official archive of the Royal Observatory in Cambridge University Library.)

This book does not pretend to be in any sense an official history of the Royal Greenwich Observatory in its Herstmonceux years. That remains to be written, building perhaps on the three-volume work by Forbes, Meadows and Howse that was published to mark the ter-centenary of the Royal Observatory in 1975. Rather, the present book is an anecdotal patchwork, published with a two-fold purpose: to set on record, before it is too late, some informal recollections of a now-vanished way of life, and to satisfy some of the curiosity of those who visit the site today and want to know what went on here in its astronomical heyday.

Herstmonceux as an observatory was unique – what other national institution in recent years has enabled people to study the heavens from the grounds of a mediaeval castle? But it exists no more. In a small way this book is an attempt to ensure that something more than just disused buildings survives to show what once made life here so special.

The authors' original contributions have been divided into sections and allocated to the relevant chapters. A little editing has been applied for the sake of continuity, to explain a number of technical terms, and to reduce duplication. The contributions of each author can be identified by superscript initials at the end of a sentence or paragraph: C – Peter Corben, H – Dorothy Hobden, J – Derek Jones, N – Bill Nicholson, Sc – Brian Scales, Se – Rosemary Selmes, Wa – Roy Wallis,

Wi – George Wilkins. Superscript numerals appear after certain other passages and refer to source notes at the end of some of the chapters. Passages not otherwise attributed have been added by the editor.

It goes without saying that I am greatly indebted to the contributors, without whom there would be no book. I am grateful also to Drs Margaret Penston and Robin Catchpole, and librarian Ingrid Howard, at the former RGO in Cambridge, for their help and for permission to use RGO photographs and quote from the observatory's *Gemini* newsletter.

Anthony Wilson   *Herstmonceux*
March 1999

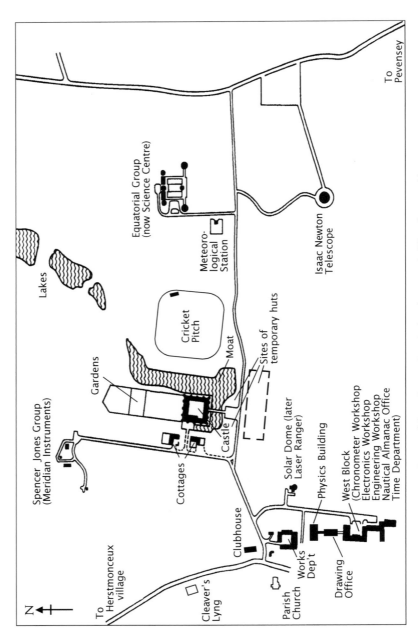

Layout of the castle grounds and other observatory buildings (not to scale).

1

Moving In

THE DECISION to move the Royal Observatory from Greenwich to Herstmonceux was taken just after the Second World War – although the need to escape from London had been recognised some years earlier.

More than two and a half centuries earlier, when Charles II founded the observatory, Greenwich had been a small village well separated from the city of London. Its main claim to fame was a royal palace on the south bank of the Thames, in whose grounds the observatory was built in 1675-6. The hill-top site, above the river mists, was a good one, and would serve the astronomers' needs well for more than two centuries.

But by the early part of the twentieth century London had expanded so much that Greenwich was enveloped. Smoke, from domestic hearths as well as factories, was becoming a serious problem. Starlight was dimmed by the murky atmosphere, and stray light from street-lamps was scattered back into the astronomers' instruments. Chemicals in the polluted air began to attack the silvered surfaces of the telescope mirrors.

In the 1920s the railway through Greenwich was electrified, playing havoc with sensitive instruments used at the observatory to monitor small changes in the Earth's magnetism. A new site for the magnetic work was set up at Abinger, 26 miles from Greenwich in the Surrey countryside. By the 1930s it was clear that the rest of the observatory's activities would also have to move out of London. But before anything could be done, the war intervened. Staffing and activities were scaled down, remaining essential activities were transferred to other sites, and valuable optical parts of the telescopes were removed for safe storage well away from London.

As soon as the war was over, the search began for a suitable new site in south-east England. From a final short-list of three, Herstmonceux in Sussex was finally chosen, and its purchase was completed in 1947.

The site was on raised ground overlooking the Pevensey levels, and some four miles from the coast. As purchased, it consisted of a castle, 370 acres of ground, and a group of temporary huts which had been used by the Hearts of Oak Benefit Society, evacuated to Herstmonceux during the war. These huts were to play an important part in the life of the observatory in its early years.

To accommodate the many different activities of the observatory, four main groups of buildings would be needed in addition to the castle. These were the Solar Observatory, the Meridian Group, the Equatorial Group (now the Science Centre) and the West Building. Post-war shortages, and other difficulties, meant that the development of the site was painfully slow. More than a decade would elapse before the observatory was fully up and running.

In 1948 King George VI gave his assent to a change of name, from *Royal Observatory, Greenwich* to *Royal Greenwich Observatory, Herstmonceux*; the 'ROG' became the 'RGO'. In the same year the Astronomer Royal, Sir Harold Spencer Jones, took up residence in the castle.

Life in a Castle

Herstmonceux Castle was some five hundred years old when it became the home of the Royal Greenwich Observatory, and has been described as 'the oldest brick building of any note still standing in England'.[1] 'It is, of course, simply a large country house and has never been the scene of any battle – it is obviously too indefensible. But there has been speculation that it might be on the site of a Roman villa – some Roman relics have been found nearby.'[N] By the 1770s the castle had fallen into disrepair, and it remained a ruined shell until restoration began in 1911. As a result, the building that the observatory took over looked mediaeval on the outside, but its interior was that of a twentieth-century country house.

'The Astronomer Royal had his quarters in the north-east corner, with his private residence above and office below. There was a small office for the secretariat, dining room, kitchen and library, with the long gallery (ballroom) divided up into offices. The Staircase Hall was used for the arrival of Father Christmas (the Astronomer Royal, Sir Harold Spencer Jones) at the children's Christmas party. While the older children had their party in the dining room, Lady Spencer Jones entertained the mothers and toddlers in her drawing room – the place

Conference delegates assemble in front of the Castle, 1982.
(RGO photograph)

was crawling! The Staircase Hall also formed a reception room for wedding parties with the main reception being held in the dining room or the Long Gallery.'ᴺ

Derek Jones worked in the castle in 1958-61: 'My office was in the South West corner on the first floor above the kitchen. I could look out over the moat and watch the jackdaws building their nests in the chestnut trees. I remember there was a time when the moat was frozen and the swans were walking around on the ice and I saw the feet slip out from under one who went down with a dull thud on to his backside.'ᴶ

'The moat supported a large number of carp and we developed a simple way of catching them. A wickerwork wastepaper basket would be lowered from the bridge until it was totally immersed. Stale bread would be dropped into it which rapidly attracted a feeding frenzy of carp. While they were busy fighting for the food the wastepaper basket would be pulled up and would usually contain a carp which could be deposited in some unsuspecting person's bath. (In those days, standard issue wastepaper baskets were made of open wicker-work. About 1980 it was decided that they were a fire-hazard and they were replaced by metal ones. I brought an old one home where I still have it. It has never burst into flames.)'[J]

Brian Scales found working in a castle something of a culture shock: 'I came to the RGO to work in the Meridian Department from the Met. Office in 1956, and night observing had not fully started. I found the change difficult at first, coming from a busy airport to a very quiet room in the castle with a Brunsviga calculator at my side and a mass of computations to perform. It seemed like a museum at the time.'[Sc] Others found that working in the country was not quiet enough: 'One staff member moved down from the middle of London to work in the new observatory. "I just cannot work here", she complained, "there is so much noise". People were surprised and asked why. "Because of all those sheep outside the window..."'[H]

But most people shared Rosemary Selmes' view that the castle was a very pleasant place to work: 'I moved into several different offices during my time at the RGO, but was lucky enough to spend my last working years there in an office, shared with two others, which had a gallery of portraits of all the past Astronomer Royals. I was also lucky in that my desk was situated in the turret, so I had windows on three sides which overlooked the moat.'[Sc]

'The castle is reputed to be haunted, by a grey lady in the present meeting room adjacent to the old chapel lecture theatre (but where was she while the castle was in ruins?) and also by a drummer boy whose ghostly drummings would warn the locals to keep out of the way of an impending smuggling run (but would not this also alert the excise men?). More recently one of the dogs was said to be occasionally reluctant to cross the bridge over the moat, but perhaps this could have been due to the presence of a cluster of fireflies, long since killed off, presumably by weedkilling chemicals.[N]

The first of the observatory's major departments to move to the Herstmonceux site was the Chronometer Department, back from its wartime base at Bradford-on-Avon. The Nautical Almanac Office (NAO) followed soon afterwards. George Wilkins, who later became

---

Ghost Story

A latter-day sighting of the spectral grey lady was reported in the Social Club's newsletter *Castle Review* in 1957. On 17 August Danny Elliot, a 26-year old stonemason on the observatory staff, described as a 'level-headed, sober young man', had left work on his motorcycle at 9.30 pm. He 'rode up to the main gate, which was open, and as he passed slowly through the entrance he bent forward, and flicked on his headlight. It was a dark, warm evening, with little wind, and the beam of light shot across the road and illuminated the gateway which provides an entrance to the church and grave-yard.'

'A woman came through the gateway and began to cross the road in front of him, and ... he stopped to let her pass. As she broke through the glare of his light, Danny noticed with surprise that she was wearing a flowing grey cloak, had short grey hair, and appeared to be moving very smoothly and swiftly, yet not running. He bent down, and with his hand jerked the gear lever of his bike into position. This operation took about two seconds, but when he looked up again, the woman was no longer there. He ... had a good look round to see where she could be – but the lady had vanished!' Certain that no normal human could have dematerialised so swiftly, the stone-mason headed for home, baffled and with 'a tingling feeling about his scalp'.[2]

---

the Superintendent (Head) of the Nautical Almanac Office, began his career at Herstmonceux a few years later:

'I joined the RGO on 1 October 1951, and during my first two years at the castle I lived in the men's hostel in one of the huts that used to stand on the area by the south courtyard.'[Wi]

'The NAO occupied huts on each side of the courtyard, and I looked out from my office across Halley Road to another wooden hut. The Chronometer Workshop, which was then extremely busy, was also in a hut. The rest of the RGO was still dispersed, with some staff at Greenwich and others at Abinger; we saw some of them on the day of the Annual General Meeting of the Association of Astronomers, as our trade union branch was then known.'[Wi]

'It was a time of economic stop-and-go, and there was little money for new buildings; one site engineer was delighted at completing one simple task – the laying of grass verges along the road to the East Gate! Another site engineer had the swimming pool cleaned out and I was able to have a swim before the water turned green again for another twenty years.'[Wi]

Bill Nicholson was another early arrival:

'After six years in the RAF as a navigator, followed by four years at St Andrews university, I arrived at Herstmonceux in 1954, just a few years after the RGO arrived there, and stayed until 1986, just before the RGO moved to Cambridge. As a professional astronomer, I consider I was extremely fortunate to work at Herstmonceux, being there for most of the time that the Royal Greenwich Observatory was based there.'[N]

'At that time a large group of temporary wooden huts provided accommodation for the Nautical Almanac Office and the Chronometer Department, plus hostel accommodation for the men (the ladies being housed in the castle, in the old servants quarters in the attics). Among the huts was a larger one used as a clubroom by the staff, complete with stage which was used for an annual pantomime, to which the local schoolchildren were invited.'[N]

'The Social and Sports Club had the use of two large rooms in the huts; one contained three table-tennis tables and a stage, while the other had a billiard table, a dartboard, and some rather dilapidated arm-chairs. The clubrooms were very heavily used at lunchtime but, as very few members of the staff had cars in those days, the evenings were usually quiet.'[Wi]

Naval Establishment

At the time of the move from Greenwich the observatory was run by the Admiralty, and this naval connection was reflected in the way things were done. When Dorothy Hobden started work as an Assistant Experimental Officer she recalls being 'made to sign the Official Secrets Act. Shortly after joining, I was sent on a three-day Induction Course in Admiralty Arch, learning about the navy and standing orders!'[H]

Derek Jones first came to Herstmonceux as a vacation student in the mid-50s. At that time 'the observatory maintained a staff of several

WAAFs in the woods

'During the war the woods behind the Equatorial Group had housed a contingent of WAAFs from the nearby radar station on the marshes, and it was reputed that also in the woods had been a secret hideout, ready for a resistance group to use in the event of a German invasion.'[N]

'There was a radar aerial at Horseye, a couple of miles from the castle, from the Battle of Britain up to about 1959. It consisted of a row of masts each about 400 feet high and was eventually demolished with explosive charges. It was replaced by a rotating aerial just to the south of Wartling village which was in turn replaced by the aerial at Fylingdales in Yorkshire. The underground installations controlling these radars were beneath the Pevensey levels. Rumour has it that they are still (1997) habitable and are the site of Wealden District Council's command post in the event of nuclear war.'[J]

messengers who visited the offices at roughly hourly intervals with material for the in-trays and to empty the out-trays. If your out-tray was empty you received a reproachful look. We were still part of the Admiralty and the messengers wore navy blue uniforms with brass buttons; one of the most frequent items they circulated were admiralty fleet orders so I knew all about the procedures to follow when de-commissioning a battle cruiser.'[J]

'Few of the staff had cars; even fewer commuted by car. The observatory had an Admiralty bus, painted navy blue with 'RN' on the side. It ran twice daily from the West Building to the castle and on to Pevensey Bay Halt at the beginning and end of work, which included Saturday morning. The bus also took the observatory cricket team to away fixtures,'[J] 'and the Social Club arranged outings to various places of interest. There was an infrequent bus service between the nearby Herstmonceux Church and Hailsham, but after a table-tennis match in Eastbourne I would have to dash for the last bus to the village and then walk down the lane back to the castle. I can even recall walking back across the marsh from Pevensey after an evening in Brighton; the sky was clear and I saw more stars and nebulae than on any other occasion.'[Wi]

'The observatory had several vans and cars and a staff of three

The Equatorial Group in construction in May 1955. The flint-knapper can be seen at work on the wall in the foreground. (RGO photograph)

drivers. In those days a large amount of the payroll was paid weekly in cash and a car was sent into Hailsham to collect the money. To provide some protection for the cashier one of the physically fit young chaps on the staff would go with him. This was a duty I shared in.'ᴶ

'Another activity was the Observatory Fire Brigade. We only practised on fine days in summer so that we could get the hoses dry afterwards. Having linked up the hose and run it to the scene of the fire we were supposed to open the valve gently so that the nozzle did not jump out of the hands of the man holding it. On one fire practice some enthusiast opened the valve so smartly that the pressure surge burst the water main on the hill. Our job was to contain a fire until a proper fire engine could come from Herstmonceux village. The Herstmonceux Fire Brigade probably consisted of a couple of full-time firemen and all the rest were retained. Up to about 1960 Herst-monceux village supported a cinema and a barber. I think the barber opened two days a week.'ᴶ

Building an Observatory

In the 1950s the observatory's building programme gradually got under way, but was soon derailed by an unforeseen setback, when the first of the new buildings failed to meet with public approval.

'In 1956 the West Building and Equatorial Group were under construction and the only functioning instrument was the Solar Dome. When this was built there had been a public outcry because it was considered to be an architectural eyesore in a beautiful part of Sussex. In reaction the Admiralty appointed a big name architect for the Equatorial group who acted with the advice of the Fine Arts Commission. The result was a poor man's Taj Mahal heavily encrusted in knapped flint, a traditional Sussex material.'ᴶ

Roy Wallis, who moved to Herstmonceux in 1957, expands on the theme: 'The Fine Arts Commission dictated much of the design of the site. They insisted that the domes blended in with the surroundings, so these were constructed of copper and coated with a substance that would accelerate the verdigris process, though everybody even then knew that a reflective white paint was the only sensible finish to buildings housing heat-sensitive telescopes. (Many years later, I installed temperature sensors all around the 26-inch telescope connected to a chart recorder. I found that the telescope's temperature, very high when we opened the dome in the evening, only would have reached the ambient about two hours after dawn! I see now – in the 1990s – that the verdigris process is starting to work!) It was also the Fine Arts Commission who dictated that the roads through the grounds should be the width of a farm track, a car and a half wide. The passing places were only achieved after a fight.'[Wa]

'The architect, Brian O'Rorke, in order to "blend in" the buildings, had specified walls clad in knapped flint. This is a traditional finish in Sussex, and I know of one of the finest examples which is a railway crossing keeper's cottage near Berwick. The flints are prepared by striking them with a sharp implement exactly along their fissure planes, just like cutting a diamond. The finished flint forms a cube of glassy blue sides. Estimates were made of flint quantities and costs were calculated, and then they looked around for a flint knapper. None were still working, but they found a retired one in Lewes, 90 years old! He was coaxed out of retirement and we enjoyed chatting to him on our way up to the West Building in the mornings. He would study a flint with practised eye and then hit it with his cleaver. Nine times out of ten he would utter a very rude word and hurl the shattered flint over his shoulder. Before long, there was a huge pile of shattered flints behind him and a very small stack of knapped flints in front of him. Most of the walls in the Equatorial Group were clad in knapped flints, but the West Building has only the bottom few feet clad in flints. Then brick takes over. The whole exercise had proved too costly.'[Wa]

'There were builders everywhere during this construction period. One lot were building the West Building, another lot the roads, a third lot the telescope domes. One morning, as we were walking up to the West Building, the shutters of one of the domes started to open. We were thrilled and stopped to watch. This was a major milestone in the history of the site. Slowly, the shutters parted. Majestically they moved along their rams. And just as majestically they fell off the top rams and leaned outwards at a crazy angle. They stayed like that until the electricians could fit stops to the top rams some weeks later.[Wa]

When it came to installing the telescopes, because the architect had designed the Equatorial Group with sudden drops, a lily pond and ha-has instead of walls, no vehicle could get near the domes. It was decided to reach through the shutter openings with a large crane. The

The North Pier of the mounting for the Yapp telescope is hoisted into Dome B. (RGO photograph)

Admiralty owned a 100 ft crane, so it was requested. "Ok" they said, "we'll send it down." Then, later, "We've lost it!" After some embarrassed searching, they found it in Northern Ireland.[Wa]

The job of assembling the telescopes fell to John Pope, then working in the Time Department. Many years later, on the occasion of his retirement in 1984, he recalled for the RGO's *Gemini* newsletter how he came to be involved:

'I was installing time service equipment in the basement of the still unfinished West Building when I received a summons from the Astronomer Royal, Sir Richard Woolley. I hurried down to the castle and standing smartly to attention in front of Sir Richard's desk heard

---

Royal Conservatory

One unforeseen consequence of the move to Sussex was the difficulty many correspondents would have in correctly addressing letters to a place called Herstmonceux. Over the years envelopes arrived addressed to Hurstmongeau, Hirst-manceaux, Hursthonceux, Hurstmagenx, Hirst Macux, Herst Man Caux, and Herstmanchox. 'Observatory' sometimes became 'Conservatory', and on one occasion the words 'Royal Greenwich Observatory' were transformed into 'Royal Suffen-wick Obstruction'.[5]

---

him say "Pope, I understand you are a qualified engineer so I want you to form an Engineering Department, recruit draughtsmen, set up a design office, and expand the workshop so that we can start building our own instruments. But in addition I have another job for you — come with me." He took me into the South Courtyard which in those days was full of large wooden huts. In one of these which housed the temporary workshop, where he introduced me to the newly-appointed workshop foreman Mick Dermody. He then escorted both of us to another hut which was full of scrap iron, or so it seemed. There were rings and tubes, odd shaped iron castings and boxes marked "Glass with care". Some had the remains of luggage labels tied to them; others had only bits of string. Sir Richard said, "These are the telescopes from Greenwich. I want you two to set them up and get them working as soon as possible so that the RGO can start observing again". With that he left us.'

'When I had recovered somewhat I said to Mick Dermody, "How on earth are we going to sort this lot out — do you have any drawings?" Mick replied, "No there aren't any drawings but I do have these," and he pulled out of his pocket some dog-eared postcards. "These postcards were sold to the public at Greenwich before the war and they show what the telescopes looked like." And so with the aid of these postcards and the memory of Jack Johnson, one of the mechanics from Greenwich, we assembled the 26-inch, 28-inch, Astrographic and Yapp telescopes in the Equatorial Group domes.'[3]

'In 1957 the Equatorial Group was approaching completion and the 30-inch reflector in Dome A was being commissioned as a

Cassegrain (a type of reflecting telescope in which the eyepiece or photographic plate is located at the bottom of the tube, near the main mirror). This 30-inch telescope had started life at Greenwich on the same mounting as the 26-inch refractor now in Dome E; the two had been presented by an eminent surgeon named Thompson. At Greenwich the 30-inch was a Newtonian reflector mounted as the counterweight to the 26-inch; it had been used to discover a faint outer satellite of Jupiter. Having two telescopes on one mounting had proved inconvenient so two separate mountings were provided at Herstmonceux.'[J]

After Hours

'Outside work, life centred round sports activities and the clubhouse, which had been built by RGO club members. There was a great community spirit. There were teams for tennis, badminton, cricket, snooker, billiards, archery, table tennis, swimming and stoolball – a Sussex game with rules similar to cricket, played mostly by mixed (friendly) or ladies (league, very serious!) teams. Enthusiastic groups wrote and produced Christmas pantomimes, and in the earlier years, when Herstmonceux village had a local Bonfire Night, a float for the procession. There were several RGO generated weddings, with receptions in the old canteen.'[H]

'There was also a strong country dancing group, joined later by a ballroom dancing section. The castle's Long Gallery was used for staff parties and dances. One I remember in particular was a New Year's Eve party which culminated with the New Year being piped in by a piper standing on one of the turrets above the main gate.'[N]

'Oxfam bread and cheese lunches were run every Wednesday from about 1967, and raised quite a bit of money, as well as being a sociable meeting point for RGO people and any visitors willing to come along.'[H]

'Cricket was played on the sports field between the castle and the Equatorial Group – not the best wicket in Sussex, but surely one of the finest settings. In 1962, a work-force led by John Hobden and containing several of the "weaker" sex, put in badly-needed drainage across the field. The Astronomer Royal (Sir Richard Woolley) was one of our staunchest and keenest members, batting at number seven. The team during the sixties was really quite strong, and included an Oxford Blue (Victor Clube) for some years. We played every Sunday, with

John Philcox, Jack Hutchins, Bill Martin, Brian Carter, Bob Wilson, Ernie Croxton, Bill Goldsmith, George Harding, Roy Wallis, John Hobden, Emrys Davies, Dave Calvert, Brent Wilson (from New Zealand, who insisted on playing in shorts, and who has sadly died recently of an asthma attack) among the stalwarts and myself as the regular scorer.'[H]

'Visiting astronomers and students who mentioned a liking for the game, were always warmly included. Afterwards, we retired to the clubhouse or the Brewers Arms in Herstmonceux.'[H] Shortly before Woolley retired, the RGO took on The World at cricket, and won. The World eleven were all of international standing – but only in astronomy, being delegates to a conference at Herstmonceux, most of whom had only learnt the rules of cricket an hour or two before the match.[4]

Setting up the facilities for these activities was a major out-of-hours task in the early years: 'When the temporary wooden huts were demolished, the club members themselves built a permanent club-house, just inside the west gate, near the church. The playing field was surveyed and levelled by the staff, and given a top dressing of mud removed from the ornamental ponds near the folly when these were being drained and cleaned out. The pavilion on the field was built using material from a small tin-roofed church (known locally as the "Tin Tabernacle") which was being rebuilt in Bexhill. Club members helped to demolish it in return for having material for the pav-ilion. The cedar cladding was provided by the contractors who were building the West Building and the Equatorial Group, and this cladding seems to have lasted over 40 years. There was also great relief when we eventually got water laid on behind the pavilion!'[N]

Roy Wallis and George Wilkins were both involved in the building of the new clubhouse:

'I was asked by Joe Bates and Henry Gill, the powerhouse of the social club, to help in club events. My reply was that people who needed to return to work in their leisure time were not quite sane and I wanted nothing to do with the club. How green I was! These were the people who had built tennis courts at Abinger with a loan from the Civil Service Sports Council matched by hard work from the members. As the West Building became complete, the need for the temporary wooden huts, including our club hut, became redund-ant. At last, the time came for the club hut to be demolished. Joe and Henry announced that we would build a new clubhouse, and they asked me to produce a design. We had converted our playing field from an alp into a level cricket pitch, so the idea wasn't completely crazy.'[Wa]

'I went to the various sports sections and asked their requirements, put all the areas together and came up with a design. It was submitted to the Admiralty architects, who moved the toilets to the rear, standardised the windows and doors and added piers at ten foot intervals and approved it. We were, of course, relying on help from the Works Department, but when it came to it, most of the builders, electricians and plumbers felt that they did enough of that during the day. There was no way they wanted to continue the same work in the evening and at weekends. One of them, Harold Rodemark, an unskilled labourer, took on the job of Works Foreman and worked all hours of daylight. Some of us laid bricks at the back of the building and cement blocks on the inside, leaving him and Joe to lay the bricks at the front.'[Wa]

'Just as the summer when we levelled the playing field had been wet and therefore perfect, so the summer we built the clubhouse was dry and therefore perfect. I remember only one wet evening when there was a heavy downpour that demolished the internal wall I had nearly completed. After it stopped, I rebuilt the wall before it got dark. When it came to the plumbing, Joe said to me, "You're going to plumb the building. Go away and learn how". I did it, but only because any time I got stuck, I just went and asked Harold what to do. He had a dozen ways to solve any problem. Unskilled?! (When it became obvious that the West Building leaked through every window as a result, I was told, of a misreading of the plans, the Works Department replaced every window, working slowly around the building. It was a wet period, and work was intermittent until Harold devised a hardboard-clad scaffold on wheels with openings at the three window levels. Work could then go on regardless of the weather.)'[Wa]

'We had the roof on by the autumn. Then some members of the local planning committee claimed that we did not have planning permission and the building must be demolished. We knew that we had approval from the Admiralty, but it took months to resolve the situation. Four of us (Joe Bates, Henry Gill, Harold Rodemark and myself) continued regularly with the fitting out until the threat had gone, and work could resume in earnest. The clubhouse was formally opened on 1 October 1960. All those who had contributed more than 200 hours received a polished wooden gavel, and their names were later listed on the panel in the RGO clubroom at Cambridge.'[Wi]

'The clubhouse provided a wide range of facilities for the staff – two full-sized snooker tables, table-tennis, badminton (but the ceiling was too low for matches), darts, a lounge and a bar. The office was soon converted into a "shop" for bulk purchases, before the days of cash-and-carry stores. The clubhouse was used mainly in the lunch-times, and only occasionally for major events as the castle ballroom was available for social events, at which the local Blue Stars band was very popular.'[Wi]

In 1965, responsibility for the observatory transferred from the Admiralty to the Science Research Council (SRC). 'The SRC tried to encourage a corporate spirit by sponsoring annual sports days, which were at first held at the Civil Service Sports Ground at Chiswick. The RGO participated strongly against the larger laboratories in cricket, netball, five-a-side football, bowls and tennis, in which I occasionally shared in winning the mens doubles or the mixed doubles trophy. After some years an indoor sports day was also held. On the first occasion I went to Runcorn as a member of the volleyball team; several of us had not played the game before, and so we were comprehensively beaten. Our table-tennis team fared much better.'[Wi]

Royal Visit

By 1958, twenty years after the idea of the move from Greenwich had first been seriously mooted, the transfer was complete. The Royal Greenwich Observatory was finally up and running at Herstmonceux. It was at this point that Derek Jones ceased to be a summer visitor and joined the full-time staff:

'I graduated from Cambridge in 1958, went to the Edinburgh Festival that summer and then arrived at Herstmonceux for my third vacation course. I stayed on at Herstmonceux after the course and was interviewed by the Civil Service Commission soon afterwards. I was successful and was appointed a Scientific Officer. The Equatorial Group was now coming into full operation. That first winter of operation was extremely trying for the maintenance staff, as the dome opening and closing mechanisms were forever failing and there were many minor teething problems to sort out.'[J]

'In 1958 we had a visit from the Duke of Edinburgh who expressed a wish to see "a normal working day" at the observatory. What he saw was most unlike a working day of any kind. Mr Allen was cleaner in the Equatorial Group and his delight was to keep the wooden floors and everything else beautifully clean and polished. He did such a wonderful job the day before the Duke's visit that we were forbidden to observe for fear of making a mark. This was much to my disgust. When the Duke came to the equatorial group I showed him Arcturus through the 28-inch by daylight and explained the workings of the telescope.'[J]

The full range of the observatory's functions at that time was summarised in a letter sent to Pearl Druce in January 1954. Miss Druce had written in to enquire about getting a job as a Scientific Assistant. She was sent a reply which detailed the qualifications required, the conditions of service and an outline of the work of the observatory, as well as the salary scales. This reply came to light 31 years later, when its content was described in *Gemini* as follows:

'At that time "the bulk of the work of the observatory (was) of the nature of long term programmes". These involved producing the Almanacs, making measurements of fundamental star positions, determining time, recording daily variations in the Sun, determining stellar distances and motions and the physical conditions in stars from their spectra, and investigating the "constantly changing parameters" of the Earth's magnetic field. Also "in the higher scientific grades (and elsewhere if appropriate) there is opportunity for individual scientific research both theoretical and practical in a very wide range of mathematical and physical problems".'

'The salary scales in 1954 are a real eye-opener. A young 16-year old man would start work as a Scientific Assistant at £215 per year, rising to £520 at age 33 if he became "Established". A woman on the other hand started at the same pay but only rose to £435 over the same number of years. What's more, these salary scales refer to the London area and there was "a sliding scale of deductions for the provinces" amounting to £10 from salaries up to £275 with correspondingly more from higher salaries. An additional allowance of (at most) 11 shillings [55 pence] per week could, however, be paid to staff receiving less than £240 and living away from home. Accommodation was provided in the observatory hostels at 45 shillings [£2.25] per week for a single room inclusive of breakfast and dinner.'

'Staff nominally worked a 44 hour week ... although this was complicated by regular night observing duties. Note, however, that "Night observing has not been an actual condition of service in the case of women but it is encouraged and several of the observatory's best observers are, in fact, women." So there!'[6]

References

1   Calvert, D and Martin, R, *A History of Herstmonceux Castle*, publ. International Study Centre, Herstmonceux Castle, Hailsham, East Sussex BN27 1RP, 1994

2   *Castle Review*, Autumn 1959

3   *Gemini* 10, April 1984, p10

4   McCrea, W H, *The Royal Greenwich Observatory*, HMSO, 1975, p52

5   *Castle Review*, August 1952, and later mentions

6   *Gemini* 13, February 1985, p11

2

Astronomers Royal

THE DIRECTOR of the Royal Observatory also held the title of Astronomer Royal (AR), until the two posts were separated in 1971. The first Astronomer Royal, in 1675, was the Revd John Flamsteed. The tenth, who presided over the observatory's move from Greenwich to Herstmonceux, was Sir Harold Spencer Jones.

'Spencer Jones was a very formal man, and I didn't really know him (other than delivering his wines and spirits with my father in my teens). When I worked at the RGO's Abinger out-station, in the middle of the Surrey woods, we tended towards informality. Sir Harold paid us an unannounced visit on a hot summer day. We were all in shorts – even the Head of Department. Sir Harold was incensed. He, of course, was in a dark suit and tie.'[Wa]

Sir Richard

When Spencer Jones retired in 1955, Richard Woolley – later Sir Richard – was appointed eleventh Astronomer Royal.

'The most memorable Astronomer Royal was probably Sir Richard Woolley. Apart from being an eminent astronomer, he loved cricket, tennis, piano playing and country dancing. People tended to be in awe of him, but he came to our wedding, since my future husband John was a fair cricketer, and so definitely "ok". Woolley also thought most highly of his regular driver, Jack Hutchins, who was also a cricketer.'[H]

'Richard Woolley became Astronomer Royal at the beginning of 1956, arriving from Australia just in time to attend the Social Club's pantomime and party. One of his first actions was to start a country dancing group, which met in the huts. Some years later there was a clash with another club function; Woolley had the office partitions in the castle's Long Gallery taken down, so that once again the castle had a ballroom. He was also keen on cricket and tennis, which was then played mainly as mixed doubles since the RGO recruited many attractive players from the local girls' grammar schools. Most games were played within the lunch-hour, but when the AR played the games went on into the middle of the afternoon.'[Wi]

'Woolley had more profound effects on other aspects of the RGO since he built up research teams at the expense of the traditional long-term observing programmes, especially those of the 'geophysical' activities, including the solar work.'[Wi]

'In his first year Woolley started the Herstmonceux conferences. To begin with they were neighbourhood conferences for astronomers in Great Britain to give progress reports on what they were doing. It soon became usual to invite a distinguished astronomer from overseas, e.g. Mikhailov from Leningrad. In the 1960s these conferences were the point of confrontation between Hoyle and his steady-state theory of the universe and the radio astronomers led by Ryle whose source number counts conflicted with the theory.'[J]

Another of Woolley's innovations affected the way the performance of members of staff was reviewed each year: 'The change from the Admiralty to SRC saw the end of a practice that was known as Staff-Side Scrutiny. This meant that each annual report on a member of staff was seen by one member of a panel of scrutineers who were elected by the staff. The aim of the scrutineers was to ensure that no member of staff was penalised by an unjust adverse report on himself (or herself) nor by the overmarking of another person.'[Wi]

'Now Woolley was keen to know about all his staff, and he carried the procedure one stage further. We had an annual meeting of all reporting officers at which they read out their draft markings and commented on the work that had been done during the year. The countersigning officers and the scrutineers were also present, and anyone could question any mark that was felt to be unjustifiably high or low in comparison with others in the same grade. I am sure that this led to a much more uniform standard of marking and it also meant that more attention was paid to the career development of staff; for example, staff were moved from one department to another in order to widen their experience, or to make better use of their talents, or even to remove a personality clash that was thought to be having an adverse effect.'[Wi]

Woolley is recalled with affection as well as awe, and many who encountered him have stories to tell ...

' ... I became active in the IPCS, the Staff Association, and attended a meeting of the Whitley Committee in the Drummers Room in the castle. (This Committee was half management and half staff representatives.) The Astronomer Royal was Chairman. We arrived ahead of the management team and settled down round the big table. Soon I felt a sharp tap on my shoulder and a stern voice said: "That's my

The Queen and Sir Richard Woolley at the Inauguration of the Isaac Newton Telescope on 1 December 1967

seat". It was Sir Richard, the Astronomer Royal, and he was dressed in topper and tails and a very imposing figure. He was due to go to Buckingham Palace after our meeting. There was a quick re-shuffle of seats. That was a definite moral advantage to the Official Side.'[Sc]

'... Sir Richard was a strange mixture. He could be very friendly or very distant. At a cricket match, he would be chatting pleasantly and then suddenly getup and walk to the other side of the ground and sit in isolation. At one match in the village a thunderstorm precipitated an early end to the game. My father was staying with us and was to pick me up from the village in the evening, so I asked Sir Richard for a lift. On the way, he asked: "Where do you want dropping off, Cleavers Lyng?" "No," I said, "I'm your next door neighbour. I live in the converted coach house. You pass it every day."'[Wa]

'... Woolley disliked visitors to the observatory. His only concession was to open the castle gardens on Tuesdays, Wednesdays and Thursdays in summer. He would allow conducted tours under strictly regulated conditions if the party could demonstrate some technical reason for coming, for example if they were surveyors or school science teachers. I took my share of guiding these parties as did many others.'[J]

The Cape Connection

'From 1959 to 1972 the Royal Observatory in Cape Town was a department of the RGO and Sir Richard, who knew South Africa well, having been born and raised there, was responsible for keeping it manned. He was finding it very difficult to send staff there as this was the time of the boycotting of South African goods. He had to put pressure on some members of staff to serve a three-year tour of duty there. Then somebody − I suspect Alan Hunter (a senior astronomer who was Deputy Director, and later Director, of the RGO) − suggested he canvass the staff for those *willing* to go. Two of us said we were, and a short while later, Peter Corben went to fill a vacancy there.'[Wa]

'A little after that, the Time and Electronics man who was running the Cape time service resigned suddenly. Sir Richard sent for me in the way he always did: the phone rang and a voice said "Come and see me". No need to ask who was speaking. He asked me to go to South Africa. I asked if I could think about it and he agreed. We lived in the cottage by the castle west entrance then, and I went home and put it to my wife. In the afternoon, I went to see Sir Richard and said yes. I then went to Eastbourne to book a voyage. We had planned a holiday on the canals and I was determined to get some sort of holiday. A sea voyage would do.' [Wa]

'The following week, the Observatory Secretary rang me and asked when I was going. "October 5th is the earliest voyage, and I've booked it," I said. "That's no good. Sir Richard will be furious," he replied. "He wants you there by the end of May when your predecessor leaves." This was the end of April. "I need more time than that," I said, "and I want to go by boat to make up for a cancelled holiday." "He'll insist on you going by air − and soon," replied the Secretary.'[Wa]

'On the Sunday, we had a home cricket match. Sir Richard and I both played in the team. "When are you going to South Africa?" he asked. "October," I replied. "Best time of year to go," he said, "the summer will be just starting. You'll enjoy it. How are you going?" "By sea on the *Cape Town Castle*." I told him. "Good idea," he said, "have you ever played deck cricket? It's great fun."[Wa]

'When I later asked for a second tour in South Africa, Woolley was delighted. Then, in 1969, Mike Candy, who was to have replaced Derek Jones in Pretoria suddenly resigned. By then, the RGO had also taken on the running of the Radcliffe Observatory in Pretoria. As part of the new South African Astronomical Observatory, with headquarters in Capetown, we had observing time at the Radcliffe, and kept a Cape observer there for the purpose. Sir Richard was visiting the Cape when he heard the news about Mike and sent for me. He asked me to work for him in Pretoria, and reluctantly I agreed.'[Wa]

'Whenever he visited Pretoria he would ask me how the observing programme was going. "How are you going to do so-and-so?" he would ask. "I'm going to do such-and-such," I would reply. "Why don't you do this-and-that?" he would say. "No, I'm going to do such-and-such.""Ok," he would say, "it's your programme." But it wasn't, it was his. But, if you were positive and had tolerably good reasons for doing something, it was fine by him. If you showed any weakness, he would start shouting at you.'[Wa]

'On one occasion Sir Richard came to dinner with Tom Lloyd Evans, Roger and Elizabeth Wood and my wife, Jill, and me. (We were the Herstmonceux-based staff at Radcliffe.) He was accompanied by Alan Hunter. Woolley told us all the news from Herstmonceux: so-and-so has got engaged to so-and-so, so-and-so and so-and-so have broken up, etc. Hunter, chuckling, said "Woolley knows a lot more about his staff than people realise. Most of the people he has told you about think he doesn't even know they exist, let alone know about their everyday lives."'[Wa]

'Some time after Sir Richard had retired from Herstmonceux, by then a widower, and also after he had retired from the directorship of the South African Observatories and been widowed for a second time (he had married the RGO's housekeeper and caterer, Mrs Marples after his first wife died), my wife and I visited Cape Town and phoned him. "Come and have lunch," he said; "Were you at the RGO in my time?" "No," I said, "you were there in my time."'[Wa]

Woolley and his third wife took us out to a well known Cape vineyard for lunch. By now he had developed a taste for good food, unlike most of his life when he regarded eating as a necessary evil to be got over as quickly as possible. We told him what was going on at Herstmonceux and Jill said something about Alec. "Who is Alec?" he asked. "Professor Boksenberg, the director." Jill replied. "And you call him Alec?" he said, "That wouldn't have been allowed in my day. Of course, you can call me 'Dick' now." We didn't.'[Wa]

Learning the Trade

'In 1956 Woolley set up a vacation course at Herstmonceux for undergraduate university students to work at the RGO for a month or so, to learn something of practical astronomy.'[J] 'At first two lots of students came for six weeks each; in later years there was just one lot for eight weeks. Boys and girls were segregated in different attic wings.'[H] '(The segregation must have come in after 1957; before then men and women had bedrooms in the same attic.)'[J]

'Each student was given a project to work on under supervision and there was a course of lectures. The first year the arrangements were fairly unstructured but regular arrangements were gradually fixed upon in subsequent years. In the summer of 1956 we worked in the castle ballroom which was subdivided into offices by wooden partitions about eight feet high but not up to the ceiling. Indiscrete remarks were easily overheard; and sometimes were. Tommy Gold, who had previously been Chief Assistant, was just leaving to be replaced by Olin Eggen.'[J] (Chief Assistants were senior astronomers in the observatory hierarchy. Some years earlier, working with Hermann Bondi and Fred Hoyle, Gold had formulated the steady-state theory of the universe.) 'He went on to be a professor in the US. We had lectures from Woolley, Eggen and Pagel in the castle chapel.'[J]

'One evening we saw a bright "star" in the sky after sunset. When Woolley was told he opened the solar dome and examined the star with the finder on the solar telescope; it proved to be a weather balloon. This wasn't the last time we were caught in this way.'[J]

'I also attended the vacation course in the summer of 1957; that summer there were two bright comets, Arend-Roland and Mrkos. On that course I met the girl who became my wife in 1960. Over the years several other students met their future spouses at Herstmonceux.'[J]

'Another 1957 student was Brian Marsden who was interested in computing the orbits of comets. He is now a world expert and head of the Minor Planet Center at Harvard. To compute the orbit of a comet, three observations are needed and they must be communicated quickly to the computer or the comet may be lost. (In those days "computer" meant a person who computed, not a machine.) To expedite communications it was customary to send the observations in encoded telegrams, a method which continued up to about 1989 when the internet became sufficiently reliable for the purpose. At one stage Brian Marsden had two observations of a comet and could not calculate the orbit for lack of a third which he anxiously awaited. To allay his anxiety we concocted a fake third observation, translated it into the international code and telegraphed it to him. This kept him happy for a day but very puzzled why the comet seemed to have such a peculiar orbit.'[J]

'Also on that course was Eric Forbes who went on to become a great authority on the history of astronomy including that of the RGO. Sadly he died some years ago. A plot was hatched to kidnap one of the students and leave him handcuffed on the Pevensey levels. The plot became convoluted by a series of betrayals which eventually resulted in a quadruple cross with Forbes being left on the levels, Nico Boersma (a Dutch student) being pushed into the moat, and the contents of my bedroom being dumped in the castle gardens. When discovered, my rolled bed-clothes were mistaken for a dead body.'[J]

'The student course was responsible for many other practical jokes. The funniest I can remember was perpetrated by some students who invited a Brighton window-cleaner to quote for cleaning the 98-inch mirror of the Isaac Newton Telescope. Another time the students wrote to the *Sun* newspaper explaining that a serious scientific journal was suffering an unexpected drop in circulation and could they have some old Page Three photographs to liven it up, please? These duly arrived. On another occasion the students timed the night-watchman's rounds over several nights, and then managed to get into the West Building and put a long line of left shoes from the entrance to the office of the Superintendent of Her Majesty's Nautical Almanac office. A bedstead and an effigy came into it somewhere too. After that, the watchmen did not have regularly-timed rounds!'[J & H]

'Another prank involved new road signs being (properly) made to replace the old ones, giving the names of current day astronomers, such as Woolley Way and Pagel Place. Unfortunately they went up on the day of a visit by the Board (a group of the great and good, who oversaw the observatory's activities), and were hastily removed which was a shame! On one occasion an old Austin car was put across the wide part in the middle of the bridge over the moat, and proved very difficult to get out again. Another time the building that housed the Isaac Newton Telescope was found one morning to have a big smile on its face.'[H]

'The student courses were held every subsequent summer, some-times two a year, and continued in Cambridge. In 1996 we held a commemorative dinner for the fortieth anniversary. Many students went on to distinguished careers in astronomy; probably the most famous is Stephen Hawking.'[J]

'Utter Bilge'

Outside the astronomical community, history remembers Sir Richard Woolley best for a single injudicious remark which he made just a year before the launch of Sputnik 1 ushered in the space age. Flying into London Airport early in 1956 to take up his appointment, the new Astronomer Royal was ambushed by reporters and asked to give his views on 'the possibilities of Space Travel. He replied that the idea was "utter bilge". This remark was to be remembered many times in subsequent years.'[J] 'It was made in a pre-interview discussion as to what they (the reporters) would like to talk about. Woolley assumed that to be "off the record".'[N]

When Sputnik 1 went up the following year, the RGO – like everyone else – was unprepared, as Roy Wallis discovered:

'On Friday October 4th 1957, the Russians launched the first space satellite, Sputnik 1. On Saturday October 5th, I cycled down to the observatory as I was due on duty in the Time Department. At the west gate, there was a crowd of people demanding to come in. I explained that no-one was there as the observatory worked a five day week (a recent change from six days). They said they wanted to ask someone about the Sputnik. I replied that our telescopes were too slow to follow a satellite and that nobody here would know any more than what was on the news anyway. "Isn't anyone else here?," they asked. "Only the Astronomer Royal," I said, "and he isn't interested in space travel." Amid grumblings, I relocked the gate and went to do my radio duty.'[Wa]

'The following Monday, the national papers screamed headlines on the front page:

GREATEST ASTRONOMICAL EVENT THIS CENTURY AND THE ROYAL OBSERVATORY SLEEPS

A spokesman from the Royal Observatory says, 'Sorry, we work a five-day week here. Nobody is interested in an artificial satellite.'

Richard Woolley's pronouncement that space travel is 'utter bilge' prompted this rejoinder from cartoonist 'Chrys' in the national press. (RGO photograph)

Sir Richard was furious. "Who was the idiot who spoke to the national reporters? I want his head on a plate. He can start looking for another job." In fear for my career, I confessed to my boss, Joy Penny, who took me down to see one of the Chief Assistants, Dr Alan Hunter. He asked me if I had known I was speaking to the press. I said I hadn't. They at no time announced who they were and I thought they were just members of the public. He said to leave it to him to sort out with Sir Richard.'[Wa]

'This was not to be the first time Alan Hunter would take my side against authority. He was a sturdy buffer between the lowliest of staff and management, and I stand with many others in gratitude to him. The result was that I heard nothing more of the matter. Sir Richard complained to the press about "anonymous investigators" and another Chief Assistant, Olin Eggen, was designated "Press Officer".'[Wa]

'Later that year we needed to find an up-to-the-minute theme for our float for Herstmonceux's Bonfire Night. Entitled "The Lost Weekend", our float depicted an astronomer in bed against a backcloth of a cobweb-covered telescope with a silver satellite revolving around the bedhead, surmounted by a Russian with a telescope and bleeping merrily, and at the foot of the bed, a cut-out of the three wise monkeys. My contribution was the Russian and the monkeys. Sir Richard was one of the three judges and refused to vote for us on the grounds that, as a large establishment with access to lots of equipment, we should not qualify. The facts that there were about half a dozen staff members involved, we had used only paper, hardboard and discarded materials, and had worked on it in the evenings were ignored. In spite of his opposition, we won first prize.'[Wa]

'Some years later, there was a lunar eclipse on a night when I was duty astronomer on the 36-inch telescope. Gordon Taylor and his team of occultation observers used every eyepiece and optical equipment on the telescope to time the occultations of stars behind the dark limb of the Moon. Afterwards, I made myself a cup of coffee before continuing my night's work and was drinking it when the phone rang.'[Wa]

'A voice said, "Have you been observing the eclipse?" I thought it was a fellow astronomer from another telescope on the site and told him I had. "Is it clear there then? It's cloudy here." "Of course," I thought, "that wasn't an internal ring. It must be an astronomer from another site." I told him that it had been, and still was clear. He said, "Could you describe it for me?" Alarm bells rang in my head. This was no astronomer.'[Wa]

"Who are you?" I asked. "This is the Central Telegraph News Agency," he said, "I want an item for tomorrow's BBC News." "I'm sorry," I said, "I'm not at liberty to speak to the press. We've got a Press Officer who will be contactable on Monday." This was Saturday night. I always get into trouble on Saturdays! The reporter tried to persuade me that a statement would be harmless and that Monday would be too late, but I stood my ground. Eventually he said, "Ok, but, completely off the record, how was the eclipse?" I replied, "Well, completely off the record, and just between you and me, it was a complete success." The item on Sunday's 9 a.m. news was "A spokesman from the Royal Observatory said that last night's eclipse of the Moon was a complete success."'[Wa]

'This time Sir Richard was amused. On his way to a country dancing event later that day he asked Mike Candy, another astronomer, "Who was the idiot with the sense of humour? Was it you?" Mike had a reputation for humorous kidding, but denied it. Mike found out some time later who the idiot was, but Sir Richard didn't.'[Wa]

Many years later Roy Wallis finally became an *official* spokesman for the RGO, when he served as Publicity and Press Officer for the observatory's organisation in La Palma, where the Isaac Newton telescope was re-established after its removal from Herstmonceux.

After Woolley

'When Sir Richard retired, the posts of Astronomer Royal and Director of the RGO were separated. I think it was because they were angling for a foreign director that the AR's title was separated from the directorship and given to a radio astronomer, Sir Martin Ryle.'[Wa]

'While they were angling for a successor to Sir Richard, Dr Alan Hunter, who had been Deputy Director for several years became Acting Director. Many of us couldn't understand why he was only "Acting". Hunter saw in the next director, Margaret Burbidge. She didn't stay very long as her husband, Geoffrey, was also an active astronomer and spent most of his time in the USA.'[Wa]

Margaret Burbidge resigned and Alan Hunter became Director in 1973. One of the Social Club's projects at that time was to return the swimming pool in the formal gardens to its former use. It had become a green, slimy lily pond, and we needed a swimming pool. We cleaned it and filled it with clean water and chemicals, but without a filter were losing the battle. When Hunter retired, we collected among the staff, as was customary, for a retirement present to remember us by. When we asked him what he wanted, he said he wanted nothing, but the club should take the money to buy a swimming pool filter and he would make up any shortfall. This we did.'[Wa]

'Hunter's successor as Director was Professor Francis Graham Smith. He was an active observer and a practical instrumentalist. This was where the rot set in (as far as informality went). People who worked closely with him called him Graham. I never really knew whether this was a forename or part of the surname, and he never gave any clues. He was very easy to work with and to talk to, and full of bright ideas about observing and detecting pulsars. By now, I had discovered a bent for instrumentation and, working with Derek Jones, was maintaining and developing the versatile People's Photometer. We would go off to Tenerife to mount it on the 60-inch Infrared Flux Collector telescope, working on Derek's programmes and on Gra-

ham's. Graham was very busy during this time negotiating with the
Spanish over the projected international observatory to be set up on
La Palma, Tenerife's smaller neighbour.'[Wa]

'I used to write and produce the RGO Christmas pantomime, and
I asked Graham to play a cameo role in it. He agreed, and I wrote a
sketch on a ship where he was a stowaway, saving all his pennies in a
bag to pay for his new observatory. After the children's performances,
I told him to prepare to lose his trousers on the staff night. He called
me a bastard, but said ok. When I removed his trousers, he had "LA
PALMA RULES, OK?" in large letters on his underpants. I'm sure
I'm the only person to have debagged a future Astronomer Royal in
public. In 1982, I went to La Palma as a support astronomer and
worked with Graham and Derek when they came to observe pulsars
with the People's Photometer.'[Wa]

'Graham had to retire at sixty and, forestalling this, resigned to take
up the directorship of Jodrell Bank, eventually to become Astronomer
Royal himself. Some time later he visited La Palma with a colleague,
and spent a night in the north of Tenerife on the way. I was in Tenerife
for a committee meeting when I received a phone call to ask me to
visit the hotel in the north to get Graham's passport as he had left it
there. I did, and it was duly sent on to him. When he next visited, I
asked him how he had got through without it and he said, "Well, you
can imagine how it was when I said I had mislaid my passport, and I
was the Astronomer Royal. There were mutterings of "Yes, and I'm
the Queen of Sheba's uncle".'[Wa]

'When Alec Boksenberg succeeded Graham Smith, I told him that
it was traditional for the director to appear in the pantomime. He said
it was not really his scene. I showed him what I had written for him,
starting, "I'm Alec Boksenberg, Director of the Royal Observatory. People call me 'Smart Alec' because I'm so clever." The scene
ends with him walking across the water, rescuing Peter Pan and Wendy
from the lake island. He read it through and said, "I'll do it!" And do
it he did – dressed in his famous Leopard Skin Underpants.'[Wa]

'Alec was always good for a laugh, even content to be the butt of
many jokes. I had quite a bit to do with him in the run-up to the
inauguration of the observatories of the Canary Islands in 1985. His
forgetfulness was well known, and he used to sidle up to Jill at a party
and whisper "Tell me everybody's names," and then go and greet them
as though he saw them every day.'[Wa]

3

Night Duty

WORKING under the open sky in the black of night is of course what makes the life of an astronomer unique. When Brian Scales arrived at Herstmonceux in 1956, he was warned of the risks of getting involved: 'It wasn't long before I was asked to do some night observing. Mr Symms, my Head of Department, said "I wouldn't start too soon, because once you do you will be observing for the rest of your working life," and he was right!'[Sc]

Roy Wallis recalls that 'it only took me about ten years to get used to working at night, and then I got by on only five hours sleep when on night schedule, compared to ten hours upwards when on day schedule!'[Wa]

Night observers were at work at Herstmonceux on every suitable night from the late 1950s to the 1980s. Their activities were concentrated in three locations: the Equatorial Group of telescopes, the Meridian (or Spencer Jones) Group of more specialised instruments, and from 1967 to 1979 the large Isaac Newton Telescope.

Inevitably it is often the long clear nights of winter that provide the clearest observing conditions. To perform best, a telescope needs to be at the same temperature as the outside air, so it is not surprising that the problem of keeping warm looms large in many observers' recollections:

'To begin with we kept warm at night by muffling up in sweaters and overcoats. Later we were issued with ex-Admiralty electric suits, powered from a 24-volt transformer and with electric slippers, gloves and hood. The Equatorial observers were given these suits but the Meridian observers had to make do with duffle coats as it was feared that the heat from the suits would disturb the accurate operation of the instruments. (I believe the electric suits were originally intended for use by deep sea divers – but other contributors to this book have different ideas, as the next few pages show.)'[J]

Observers who had no transport of their own could sleep in the castle but those who owned cars would sometimes come by car. These were often old or second hand and were unreliable to start after a cold night's observing. Some people would park their cars behind B Dome facing down the hill so that it was easy to push start them. One chap had a little oil fire which he used to leave burning under the engine.'[J]

'It could be extremely cold – even at Herstmonceux. Working on the Danjon Astrolabe, which was housed in a wooden hut on a raised floor, meant sitting with head and shoulders in the open air. The Admiralty had a batch of electrically-heated suits made for the deck crews of aircraft carriers. It was quickly found that, as they were nylon, in the event of a fire, they would melt and cocoon the unfortunate occupant *en croute*. Each dome was fitted with a 24v transformer, and we were issued with the redundant suits. On the Astrolabe, I used to throw a blanket over my shoulders to keep the heat in and huddle down for warmth.'[Wa]

'One night I smelt burning. I carefully walked around all the equipment sniffing for the tell-tale smoke – nothing. I sniffed outside – nothing. Eventually, I opened my blanket and the smell intensified. When I unbuttoned my collar, smoke drifted out. I felt myself all over, and when I put my hand on the front of my thigh, felt a stab of pain. I hurriedly climbed out of the suit and found my trousers on fire. A smouldering hole, three inches across ventilated my leg. I never got compensation for the trousers. The Secretary ruled that it was normal wear-and-tear.'[Wa]

'The winter of 1963 was exceptionally cold, with snow from New Year's Eve, through January and February to early March, rather than the normal few days here and there. The moat was completely frozen over – the only time when it was entirely safe to go skating on it. I had been home to Northern Ireland for Christmas, and as planes were grounded, had to come back by boat to Liverpool and train to London and Eastbourne, where I arrived 24 hours later. I was anxious to get to work, but found I then couldn't get across the marsh to Herstmonceux. In the ensuing white frosty weeks, we several times had to walk some of it. (Nowadays, people might not be so industrious!)'[H]

'Observing at Herstmonceux provided a great variety of experiences. I have known the wind blow so hard that it snatched the door of Dome F out of my hand as I was leaving. Another time I was on the catwalk of Dome B and saw a line squall approaching over the fields at about fifteen miles an hour. A line squall can be identified as a line of low black cloud with rags of cloud hanging down below it. The air in front is comparatively calm but it sets every tree tossing as it is reached. Another night it was bright Moon and very calm. The whole of the Pevensey levels was covered in a sheet of mist with only Horseye rising through it.'[J]

'The "Hurricane", or Great Storm, of October 1987 came up in the night through Hampshire, Sussex and Kent, losing its ferocity further north. But it was strong coming along the coast and through Eastbourne and Herstmonceux, cutting down swathes of trees and causing damage in its path. I tried about four different ways to get into work, eventually making it late morning. Many didn't make it, as there were trees down across roads everywhere.'[H]

'Incidentally, most of the trees on the estate were planted on the initiative of Sir Richard Woolley. This was not just an early foray into ecological planning, but had the more practical intention of creating equable surroundings for the telescopes. I forget the total number of trees he and the head groundsman planted – several thousand – but one slightly unfortunate result was that the view of the castle was blotted out on the eastern side by eucalyptus trees, a reminder for Sir Richard of his time in Australia.'[N] '(My recollection is that Jack Pike used to claim that he and Cyril had planted 250,000 trees.)'[Wi]

'One of the "characters" of the early days in Herstmonceux was Jack Pike, the forester, who was greatly respected by Sir Richard. Pike claimed to have been to Oxford, though this was probably only true in the geographical sense! There were some lovely old trees in the grounds, which were checked for safety by a surveyor from Chatham who pronounced them safe. However Jack Pike told the AR that one of them did not look safe to him, and would probably come down in the next storm. Two weeks later there was a great thunderstorm, and the tree fell across the bridle path. So much for the experts...'[H]

Observing was possible from Herstmonceux roughly one night in three, averaged round the year. The others were marred by moonlight, cloud, and other adverse conditions: 'Herstmonceux is a low site near a marsh and observations were often stopped by fog and when dew covered the front surface of the object glass, or more seriously dewed up the surfaces between the two components of the objective.'[C] 'To avoid this the refractors were provided with "dewcaps" – cylinders projecting forward by around three feet from the object lens. When these failed to work we used hair dryers to remove the dew.'[J]

The Equatorial Group

The Equatorial Group was so named because all its instruments have 'equatorial mounts', an arrangement that makes it easier for them to turn to follow the stars as they move across the sky during the night. Telescopes in this group included refractors, whose main optical component is a large converging lens called the objective, and reflectors, whose principal component is a large curved mirror. The diameter of the objective lens or mirror determines, amongst other things, how much light it gathers and hence how good it is at detecting very faint objects in the sky.

Prominent among the telescopes of the Equatorial Group were (and still are) the 26-inch 'Thompson' refractor, the 13-inch 'Astrographic' refractor, and the 36-inch 'Yapp' reflector. All three were designed for taking photographs, rather then for looking through. A secondary telescope called the guider is attached to the side of each instrument, and the observer looked through this to check that the main instrument was correctly aligned.

'Basically, astronomy falls into two fields, astrophysics and astrometry. Astrophysics determines the properties of individual stars, what they are made of, how big they are, and so on, but could really examine only one star at a time. Astrometry, on the other hand, deals with the positions of the stars, how far away they are and how they are moving, and usually deals with many stars at a time, particularly when photographs are used to determine the positions in the sky. This then leads to practical applications of astronomy in time-keeping and navigation. Indeed it was with this latter aspect in mind that the observatory was originally set up in Greenwich.'[N]

The observing programmes for the Equatorial Group included both aspects − astrometry mainly done with the refracting telescopes, and astrophysics with the reflectors. Peter Corben was one of the astronomers who moved to Herstmonceux from Greenwich when the old Royal Observatory finally closed in the Spring of 1957, and was soon back at work in the Equatorial Group:

'By 1958 the Equatorially mounted telescopes, previously at Greenwich, had been re-erected at Herstmonceux. The Equatorial Group was built with six domes − three for reflectors on the north side and three for refractors on the south side. The buildings were designed by an architect with little input from the Royal Observatory

The six domes of the Equatorial Group, seen from the Isaac Newton Telescope. (RGO photograph, 1974)

staff. In consequence the group has features which are less than ideal for astronomical telescopes. Firstly, it is important in an observatory for observers to preserve their dark adaption (the extra sensitivity that the eye develops after a period in darkness), so they like to walk around using the minimum of light. The Equatorial Group has numerous steps, a large pond and what were then unfenced ha-has. At least one student fell into the pond after coming down the steps from the Astrographic dome. (The problem of dark adaption is no longer so serious in a modern dome, since the guiding of a telescope is now normally done using an autoguider or from a control room using a TV image.) Secondly, the reflector domes in the Equatorial Group are joined together in one building which makes them very slow to lose daytime heat affecting the seeing at night, and thirdly, the domes are copper-covered instead of being finished with heat reflecting white paint.'[C]

'I started my observing in the Equatorial Group on the 26-inch

refractor. The chief programme on this instrument was the deter-
mination of stellar distances or "parallax". This was the continuation
of a programme from Greenwich and is the only direct method of
finding the distance of stars. The method is only suitable for telescopes
which have a long focal length and so have a large scale at the focal
plane. Stars which were thought to be close because they had a large
motion relative to other stars, were chosen for the programme. The
26-inch telescope objective is a doublet, figured to give a sharp image
in blue light and so it is used photographically.'[C] (A doublet is a pair
of lenses close together and made from different types of glass. The
combination can overcome some of the problems of getting a clear
image with a single lens.)

'Photographic glass plates coated with a fine-grained blue-sensitive
emulsion were used for the programme. The fine grain ensures that
the star image can be measured with the highest accuracy. A photograph of the star whose distance was required was taken and repeated
six months later when the earth was on the opposite side of its orbit
round the sun. The movement of the star, relative to other stars on
the same plate, gave the distance of the star. In practice the measurement is very small. Some 30 plates were usually needed over a period
of several years to separate out the parallax and proper motion (the
slow steady motion of the star across the sky) of the selected star. The
magnitude of the central star was reduced to the same as the average
of the reference stars by means of a rotating sector with a suitable cut-
out so that all the star images were of similar brightness, and hence of
similar size, and could be measured more accurately. Observations
were always made near the meridian and the objective lens was never
touched apart from careful cleaning.'[C]

'The observer guided the telescope using an eye-piece on the 13-
inch Merz telescope (the guider) which is securely fixed to the
26-inch. If the central star was too faint for use as a guide the observer
could move the eye-piece across the field of view to pick up a suitable
star at predetermined off-sets from the central star. Towards the end
of the 1960s an autoguider was fitted to the Merz. This measured the
error if the star image drifted away from the centre and applied the
appropriate corrections to the telescope.'[C]

Speed of operation was important with the 26-inch, so its dome
was given a special feature: 'The dome was given a rising floor because
many exposures were only three minutes in duration and it was

---

Record Breaker

When E. D. Clements retired in 1982 it was recorded in *Gemini* that in the course of his 26 years at the RGO he set a number of 'observing records which are unlikely to be broken. In one night of observing with the Transit Circle he and Brian Scales together observed the record number of 540 stars, while with the 26-inch telescope he broke the record 14 times and set the record of 43 plates obtained by a single observer in one night.'[1]

---

necessary to move from one field to another as quickly as possible to observe as many stars as possible while the sun was well away from the meridian.'[C] This refinement was needed because the 26–inch telescope is 22 feet in length. With such a long telescope, the observer's position can move a long way when the instrument is turned to point in a new direction. The rising floor makes it possible for the observer easily to reach the correct position whichever way the telescope is pointing.

'The telescope was also used for photometry (measuring the brightness of objects) using different colour filters and appropriate photographic emulsions.'[C]

Rosemary Selmes also has vivid memories of her work as a young astronomer using the 26–inch:

'The Thompson telescope is over 100 years old and was used as a giant camera, taking photographs of distant stars and galaxies invisible to the naked eye. It is a large telescope, 22 feet in length and with a lens at the top measuring 26 inches in diameter. Such a long telescope needs a dome with a moving floor, in order to manoeuvre it to point to any position in the sky without hitting the floor, but still remain in reach without the astronomer using it having to climb ladders.'[Se]

'Imagine a cold winter night with a heavy frost on the ground. In the dome there is no heating and the only light comes from the small red lights by the clocks. I am on my own. I am only 17 years old, and this is my first night's observing alone, following my basic training. I am also quite short, so moving the telescope around is quite a task. All telescope manoeuvring has to be done in the dark. Great

care is needed to ensure that the telescope does not run into the floor
or any furniture, and you need to avoid bumping into anything. It is
very difficult to keep warm and everything you touch is metal. Gloves
are okay until you need to do any delicate jobs, then they have to come
off.'[Se]

'In order to take long exposure photos, a good tracking system is
needed, or we would have lots of photos of sausage-shaped stars! The
26-inch has a motor attachment to track the stars across the sky. There
are two problems with this particular drive system. Firstly, it only runs
for two hours before it needs rewinding, so you have to be sure you
have enough time left on the drive before you start your exposure,
and secondly, at some angles, it is not quite accurate enough to give
clear images. This means that all fine adjustments needed to be done
by hand. This is where the glamorous idea of observing disappears.'[Se]

'When the telescope is pointing at the correct area of sky and the
photographic plate is in position, the observer then has to use the
eyepiece to find a bright star in the field of view. This star is then
centred on some fine cross-wires in the eyepiece and the observer has
to spend the next hour either sitting or lying under the telescope with
one eye glued to the eyepiece ensuring that the star does not move
off the cross-wires. If the star begins to move to one side, fine
adjustment screws are used to bring it back to the centre. You can
imagine that there are occasions when the observer falls asleep.'[Se]

'Apart from what is happening to the images on the plate, you have
to imagine the direction the observing end of the telescope is moving
as it follows the target star across the sky, usually downwards. Getting
trapped beneath the telescope can result in a very painful experience,
particularly if the floor controls are almost out of reach. In the dark,
it is extremely difficult to find the one button that moves the floor
downwards. Pressing the wrong one creates even more pain!'[Se]

'Having finished the exposure of one photograph, it is time to
quickly reset the telescope ready for the next picture. So many nights
have cloudy skies that whenever the sky is clear we have to take
advantage of that and produce as many photographs as possible. This
means there is no time between exposures for taking a look at the
spectacular rings of Saturn or at Jupiter's moons.'[Se]

'There is however plenty of time for gazing up at the night sky
and contemplating. Anyone who ever takes the time to study the stars
will know what I mean when I say: "There is just something mystical

The 26-inch Thompson refractor in Dome E of the Equatorial Group. (RGO photograph, 1974)

up there". You are seeing things which happened back in history. In other words, it takes so long for the light to travel from the star, or whatever you are watching, to your eye, that by the time you see what is happening to that star, hundreds, maybe millions of years have passed and the star may no longer even be there. Being an astronomer made me realise just how small and insignificant we really are.'[Se]

'Returning to more practical aspects of observing, consider just how cold it becomes in the dome. Imagine a beautiful clear sky on a mid-winter night with a heavy frost on the ground, and remember that an observer working during the late 1960s spent a lot of time lying under the telescope and not moving around. I know that some of the men did wear the electrically-heated ex-RAF flying suits that were issued, but imagine how these fitted the girls. Several of us were quite short and virtually disappeared inside the suits, and it would have been almost impossible to move around. We decided it was better just to stick with our many layers of assorted clothing.'[Se]

'One hazardous aspect of working on the 26-inch was the weather checking. If you spend your time just looking at a small area of sky through the open shutters of the dome, you are unaware of the weather changes in the rest of the sky. If a rain cloud approaches from your blind side, the first thing you know about it is when you get wet. This obviously needs to be avoided as it would cause a lot of damage to the telescope. Regular checks of the weather from outside the dome are needed, and this is where the danger arises. Some observers would use different telescopes on different nights. You have to remember that the 26-inch dome is the only one with a rising floor. This means that instead of just popping out of the door, you have to carefully climb down the vertical stairs first. Needless to say more than one person had a nasty fall. Another hazard came from working in D Dome – the one by the pond. The pond is immediately in front of the dome's main door, and the most direct route to the main building in order to go home at the end of a shift, is straight through the middle of the pond! There was more than one occasion when someone carrying their photographic plates back to the main darkroom, went straight into the pond and destroyed a whole night's precious work.'[Se]

'When I started at Herstmonceux, Sir Richard Woolley was the Astronomer Royal, and he and Lady Woolley were living in the castle. There was a time when Lady Woolley became very concerned about young girls working at night without male protection, and she insisted that there should be at least one man on duty on each observing shift. She later got to hear that if the skies clouded during the middle of the night, everyone spent the rest of the night in the rest room, "eating, drinking and socialising", whilst waiting for the

Fly in the ointment

When disaster struck in the Equatorial Group in 1987, astronomer Michael Penston reported the event for *Gemini*: 'Spider eats fly is no news but, when a dead fly fell down the tube of the Thompson 26-inch refractor at Herstmonceux and broke the spider-web cross hairs in the guider, your ever-eager *Gemini* reporter rushed to the scene. It was too late to interview either spider or fly of course but the incident revealed a loss of technological capability at RGO where no-one remembers how to train a spider to make a cross-shaped web! A temporary replacement with human hair showed just how thick the donor was compared to the very sharp spiders of yesteryear.'

This news item caught the public imagination and was reported in the press in the USA, Canada, Australia, Germany and Eire, as well as in the papers and on radio and television in Britain. Letters and offers of help were received from web-collectors around the world. It also transpired that the original report had been wrong. The following issue of *Gemini* offered 'apologies to those members of RGO's present staff who do know how to collect web in the autumn, store it and later install it in the eyepieces of telescopes.... For the moment however the problem has been solved with a man-made fibre.'[3] The damaged eyepiece had been repaired with a nylon filament 20 micrometres thick and 'the telescope has been working well since'.

weather to improve. This put her in somewhat of a dilemma and she was unable to decide which situation put the girls most at risk!'[Sc]

'At the end of my night shift, being without a car, I had to sleep in the castle dormitories. This meant a long lonely walk from the rear of the Equatorial Group down the hill, past the woods and the castle drawbridge, and round through the long avenue of chestnuts to the far entrance of the castle. Then a long tired climb up several flights of stairs to the small bedrooms on the top floor. The plumbing in a five hundred-year old castle leaves much to be desired, and the bedrooms had horizontal pipes running above the ceilings carrying air bubbles, which sounded like mice running to and fro.'[Sc]

'My job at Herstmonceux was as a Scientific Assistant working on the quasar research team. I was very lucky to have completed my basic training at the time when quasars had just been discovered – objects which looked like stars but are as bright (we now know) as whole galaxies. At that time no-one knew quite what quasars were, and so two research fellows were sent down to the RGO from Cambridge to start a research project using optical observations. They needed an assistant, and I was available. I spent seven happy years working on the quasar project, first with Michael Penston and Russell Cannon, and later with Keith Tritton. I feel very privileged to have been part of this team, and I love going back to Herstmonceux today, to the Science Centre, and sharing some of my experiences with visitors there.'[Sc]

Derek Jones was also involved with a number of research programmes using other telescopes in the Equatorial Group, including a project to 'weigh' stars:

'My first responsibility was the 28-inch refractor in Dome F (now the home of the 38-inch "Congo Schmidt" telescope). Woolley, Symms, Candy and I were pursuing a programme of observing the position angles and separations of double stars. These data are necessary to calculate the orbits which the stars follow in their motion around each other. These orbits in turn lead to the masses of the stars. I am the only survivor of the quartet and have been helping the National Maritime Museum with the 28-inch which was returned to Greenwich in 1971 and is on show to the public in the Old Observatory there.'[J]

'I also had a programme on the 36-inch reflector in Dome B, measuring the radial velocities of stars in an association of young stars. (Radial velocity is the component of a star's speed along a direction towards or away from the Earth.) I was trying to discover whether or not the cluster was expanding.'[J]

'When O. J. Eggen arrived at Herstmonceux he wanted a small telescope suitable for photoelectric photometry – measuring the brightness of objects in the sky using an electronic detector, rather than photography. There was no telescope available so he arranged to borrow the Isaac Roberts Telescope from the Science Museum and mount it in Dome C. This nineteenth-century instrument had been used by Isaac Roberts for his epoch-making photographs of nebulae – the first photographs of what we now know to be galaxies far beyond

Blind eye

When Phil Cottrell retired in 1983, one of the memories he particularly cherished from 18½ years at the RGO was being presented to the Queen at Greenwich on the occasion of the re-inauguration there of the 28-inch telescope after it had been returned from Herstmonceux. 'The Queen was later invited to look through the telescope. However, it was only when the Duke of Edinburgh complained that he couldn't see anything that it was realised that the end cover of the telescope was still on! The Queen had been too diplomatic to say anything.'[5]

the limits of our own Milky Way galaxy. The telescope wasn't a great success for photoelectric work but I did observe with it for a few nights. It wasn't driven by an electric motor as a modern telescope is, but by falling weights with a centrifugal governor. The weights were above the level of the observing floor in Dome C when wound up but gradually fell to the ground floor in operation. One night I was observing when the weights reached their lower limit so I began to wind them up with the ratchet mechanism. Just as they reached the top the linkage of the wire rope parted and there was a distinct moment of silence before an enormous crash as the weights hit the floor.'[J]

'I also observed on the 13-inch astrographic refractor which used the same 16 cm × 16 cm plates but produced photographs at half of the scale of the 26-inch. Most of the work on this telescope was concerned with repeating plates of the International Chart of the sky, to determine the proper motions of stars for which no second epoch plates had been obtained.'[C]

'A group of night watchmen patrolled all parts of the observatory by night with a timing box which recorded their visits to each site. When you were observing you would get two visits a night but you wouldn't know when because their route and times were changed randomly every night to frustrate any planned intrusion. The watchmen had to make their patrols on foot whatever the weather and Woolley was furious when he discovered some miscreant using a car.'[J]

'At the end of each night's observing the observing books were placed in a tray beneath B Dome. In the morning they would be taken by messenger to Dr Hunter who would provide an abstract for the Astronomer Royal. The Dome F book would then come to me and I would maintain a card catalogue of the observations. The other books went to people responsible for the other telescopes. The books were returned by messenger to the Equatorial Group before evening.'[J]

Astronomers were not the only creatures active around the Herstmonceux estate at night: 'In the 1960s it was quite common to hear the nightingale.'[J] 'Fireflies were also to be seen (a quite startlingly bright bluish green) near the telescopes at night'.[N]

'One night I was working on the 26-inch telescope when I heard blood-chilling screams. I didn't often get scared alone at night, but this time I was petrified. It sounded as if at least two people were being raped and murdered! I just couldn't stay in the dome and wait for the assassin, so I went out on to the balcony. I shone my torch on to the roadside verge, and there were two young badgers rolling about and screaming. I told my colleague, Clem, who was a badger watcher, about it and he was very envious. "I have read about the badger screams, but I've never heard them. They do it when they are having fun," he said.'[Wa]

Observing – taking the photographs – was only the first stage of the work. The glass plates then had to be developed and analysed. Analysis often meant many hours work with a microscope measuring off the precise positions of the images of individual stars, a process which was later automated:

'It was the custom to develop the spectroscopic plates from Dome B in the dark-room immediately to the East of the telescope. The direct plates from Dome E and D were taken to the West Building by the messengers where they were developed. The boxes were returned to the Equatorial group by evening with the observing books.'[J]

'Also housed in the basement of the West Building was a large measuring machine, weighing some four tons and settled on the bedrock for stability. It was used to measure the positions of the stellar images on the glass plates taken with the telescopes. It could measure at the rate of about 900 images an hour, to an accuracy of a micron (a thousandth of a millimetre). Among several projects, its main use was in measuring positions of stars on 5820 plates covering the whole of the sky south of the equator in a series of overlapping photographs. The resulting measures were collated like a large jigsaw, leading to the compilation of a large catalogue of star positions in the southern hemisphere.'[N]

The Meridian Group

The instruments in the Meridian Group were used for precise measurement of the positions of stars and planets. They were the Danjon Astrolabe, the Reversible Transit Circle (RTC) and the Photographic Zenith Tube (PZT), which was the most modern and precise instrument, used for the determination of time in terms of the exact moment when particular stars pass overhead. It was capable of timing the exact moment when a star passed overhead to within a few thousandths of a second. From 1961 the operation of this instrument was made largely automatic.

The Meridian Group stood on rising ground to the North–West of the castle, about half a mile from the Equatorial Group. The work done there was routine, but central to the original function of the Royal Observatory, the provision of precise data – mainly star maps

---

'Up at the PZT'

Conditions for the operator of the Photographic Zenith Tube were less spartan than elsewhere, as Philip Rudd explained in this poem, published in *Gemini* in 1982:

'I looked out at nine thirty
The stars were shining bright
I said a rude four letter word
And went to start the bike.

An observer's life, so the story
    goes
Is the only life for me
But I began to have my doubts
On the way to the PZT.

I got the brutie a'working
Then lay back and lit my pipe
I watched the first star pass OK
A trend, I hoped, for the night.

At least it's nice and warm in here
It really is a cinch
Which is a damn sight more than
    can be said
For the cold dark 26-inch.

It's not that I am knocking it
It hardly was a bore
After all how could anyone tire
Of that amazing rising floor.

After this declaration
I'm sure you'll plainly see
It's not really such a bad old life
Up at the PZT.'[4]

and tables – that would allow people to use the sky to determine their position anywhere on the Earth. One instrument, the Reversible Transit Circle, was involved in the 1950s in a programme to observe and measure the position (at least twice each) of no fewer than 13,500 individual stars:

'The building housing the RTC was a rectangular barn-like building. The two halves of the roof slid back lengthways and part of the ends of the building dropped down. Consequently, there was very little protection for the two operators and one would get extremely cold in the winter. At first, we provided our own warm clothing and some bizarre outfits were concocted in order to keep warm. Eventually we were provided with Admiralty clothing like seamen's sweaters and electrically heated submarine suits. You couldn't use the heating in these suits for fear of upsetting the observations. I settled for Kapok filled jacket and trousers called naval arctic suits.'[Sc]

'We worked in pairs on the RTC, one observer and one "mic reader". The term mic reader was inherited from Greenwich when that person had to set and read the micrometers for the declination circle every observation. At Herstmonceux these micrometers were replaced by 35 mm cameras but the term persisted.'[Sc]

'My observing partner, Clem, (E. D. Clements) used to bring a bottle of milk for refreshment and to his consternation it froze one night while we were observing, protruding from the bottle in a solid lump. On many occasions my breath froze on the face plate of the telescope and icicles formed.'[Sc]

Brian Scales also worked with the Photographic Zenith Tube at a time when it was prone to teething troubles: 'I soon started night duties on the PZT and, being new technology for that era, it was very temperamental. There were many problems at first. One frequent fault was that the plate carriage used to drive too far on its traverse, would jam and then a warning buzzer would sound. One particular night the carriage was particularly rebellious and my notes in the observing book read:

> Buzzer ... observation
> Buzzer again ... observation
> Buzzer again ... observation
> Buzzer yet again
> Buzzer, buzzer, buzzer.

Next morning when I arrived at the office, a little late after my frustrating night duty, I was confronted by most of my colleagues chortling with glee while poring over my observing notes. "What's so funny?" I said. The reply was "Your z's look like g's". My handwriting improved after that![Sc]

The Isaac Newton Telescope

For a brief period in the 1960s and '70s, and after a long gestation period, Herstmonceux was host to the largest telescope in the world outside America and Russia, the 98-inch Isaac Newton reflector. 'This had been first projected in 1946 in the first flush of postwar reconstruction after World War II. The primary mirror had been given to the project by an American foundation. It was a 98-inch disc of a substance similar to Pyrex which had been cast as an experiment during the work on the Palomar 200-inch. The Isaac newton Telescope (INT) was repeatedly postponed because of successive financial crises and its design was in the hands of a group of astronomers with little management experience.'[J]

'When Woolley arrived in 1956 he found that the design was a Schmidt camera with many innovative features which might well be unreliable under operational conditions. He succeeded in getting the project moving again and changed the design to a much more conventional reflector with Prime, Cassegrain and Coudé foci – terms which refer to the different configurations in which such a telescope can be used. The INT was in fact similar to the 74-inch telescope at Mount Stromlo observatory in Australia where Woolley had been director before coming to Herstmonceux.'[J]

'During the mid-1960s we watched with curiosity the building of the dome for the Isaac Newton Telescope and then the lifting-in of the parts of the telescope.'[Wi] The main mirror for the telescope arrived at a bad moment, in the middle of a cricket match: 'One Sunday a game was interrupted when a lorry arrived, with the driver coming over to ask where to deliver his load, the 98-inch mirror for the Isaac Newton telescope. One can imagine the fanfare that such a delivery would have aroused in the USA, and indeed did for their 100-inch and 200-inch mirrors.'[N]

'Some of us were invited to the Long Gallery when the Queen came on a foggy day at the end of 1967 to inaugurate the telescope. It

The 98-inch Isaac Newton telescope. The operator can just be seen in a seat attached to the D-ring at the back of the telescope. (RGO photograph)

had, however, been made clear to us that this was not to be "an RGO telescope", but was for the use of university astronomers.'[Wi]

The Isaac Newton telescope was so large that while it was at Herstmonceux the observer could actually ride inside it while it was in operation: 'I also worked in the Prime Focus cage of the 98-inch taking plates for parallax observations. But when the 98-inch was moved to La Palma with a new 100-inch mirror, the prime focus was operated using an autoguider in place of the observer.'[C]

'Observing on the Isaac Newton telescope was a new experience for most of us. Here, at last, was a telescope with a closed control room – sheer luxury to an astronomer used to being subjected to the cold, although the Director at the time told one of the staff that astronomers should be cold and uncomfortable or they were not doing their job! True, the new generation of computer-controlled detectors hadn't yet arrived, and the observer still had to cling to the back of the telescope with eye glued to the eyepiece during the observation, but at the end of each object, we could go into the control room and warm up.'[Wa]

'If there were two of us, we could take it in turns to ride the telescope – literally, because there was a D-shaped ring attached to the back of the telescope with a seat mounted on it, and the observer could drive the seat from side to side on the D-ring, and the ring itself backwards and forwards. This sometimes left the observer in a very uncomfortable position with head horizontal and a pain in the neck. One of the modifications I was responsible for was the addition of a headrest bolted on the telescope for cranial support.'[Wa]

'Because the telescope was driven remotely, a new category of staff was employed – the Night Assistant. Astronomers who had worked at more technically up-to-date sites were used to these, but I wasn't. The Night Assistant would reset the telescope to the next object, keep the log and watch for the security of telescope and observer. A model of the telescope with a transparent dome sat on the Night Assistant's desk and was coupled to the actual dome and telescope. The Night Assistant could see from this that the telescope was looking out through the slit and not obscured by the dome, and he could keep the two in line.'[Wa]

'There were three Night Assistants. One was a farmer by day, one had been in the pop scene and the third, a keen amateur astronomer, had tried more than once to join the staff without success. The two latter did join the staff after the INT was moved to the Canaries, and both went with it, one as an electronics technician and the other as a support astronomer.'[Wa]

'Of course, the weather wasn't always cold – only mostly. I do remember one really hot night in 1976, when I worked all night in just a pair of shorts, and even then, while the instrument was integrating the light from each star, I lay on the steel floor trying to lose heat from my bare skin.'[Wa]

'If, as I was, you were also developing instruments to be mounted on the INT, and had to spend parts of the day testing and calibrating, you sometimes looked over your shoulder and, with a start, saw a dozen pairs of eyes watching you. This was because the public were allowed into the viewing gallery to view the telescope. They entered by a separate door, mounted a spiral staircase (as a scientist, I should call it a helical staircase) and stood behind a double-glazed window through which they could see the observing floor. If you strained your ears, you could hear muffled talking, but normally you were unaware of their presence until you inadvertently made eye contact.'[Wa]

'One curious effect experienced by several observers on the INT was focussed sound. To while away the long nights, most observers took a portable radio with them and listened to music all night. Occasionally, observing a northern object would mean that the dome slit of the INT was exactly directed towards one of the Equatorial Group telescopes at the same time as that telescope was aiming its dome directly at the INT. The other observer would suddenly be overwhelmed by your music and you by his. The two domes would act as concave reflectors and the two observers would be at their foci and receive a highly amplified sound signal.'[Wa]

'Do I miss these interesting nights? Well, perhaps, but sometimes I wake in the night, remember them, and then snuggle down into bed.'[Wa]

References

1   *Gemini* 4, November 1982, p8

2   *Gemini* 17, September 1987, p27

3   *Gemini* 18, January 1988, p21

4   *Gemini* 3, September 1982, p5

5   *Gemini* 8, October 1983, p9

4

Time and Tables

BY NO MEANS all the work of the observatory was done at night. The majority of the staff, who by 1974 numbered more than 200, in fact worked fairly normal hours. Substantial effort went into activities such as running the Time Service and producing The Nautical Almanac – although both these daytime activities relied on the results obtained by the night observers.

Measuring Time

The Royal Observatory is perhaps best known for its role as custodian of the nation's time, and as the reference point for time zones throughout the world. For many years, Greenwich Mean Time (GMT) was determined by the rotation of the Earth, midday GMT relating to the moment when the sun is due south in the sky at Greenwich. For practical reasons, however, it is easier to use stars rather than the sun to measure time, and this was done with instruments of the sort found in the Meridian Group at Herstmonceux. The function of the RGO's Time Department was to maintain a public time service for Britain, and to make this available to outside users, for example by generating the broadcast "pips", which for many years were produced at Herstmonceux and sent to the BBC by landline.

Roy Wallis joined the Time Department when it was still in Abinger, and moved to Herstmonceux before the clocks did:

'During the Second World War, the Royal Observatory was split up and spread around the country. The Time Department (in those days considered an essential service to the war effort) was moved to Abinger in Surrey, with a few clocks and a skeleton staff remaining at Greenwich, and set up in the woods below Leith Hill and that's where I began my career.'[Wa]

'I moved to Herstmonceux ahead of the rest of the Department, as my wife and I had chosen the first of the council houses to be completed in the estate. This was in 1957, and the observatory had

'On the proposed move from Greenwich'

When Dr D. S. Perfect retired from the Time Department in
1954 he donated to the RGO a number of poems, including
this sonnet on the theme of Time, adapted from
Shakespeare's Sonnet 123, which starts 'No, Time, thou shalt
not boast that I do change!' Perfect's version is dated 1946,
the year when the move to Herstmonceux was announced:

No, Time, thou shalt not boast we do not change,
These piers and pivots built with newer might
For Thee, are something novel, something strange
Not merely relics of a former site:
Our dots are brief and we no more admire
What thou dost foist upon us that is old,
We rather make new-born our new desire
And dream of ways we've not before been told.
Our registers indeed may truth deny,
Thy wanderings in the present as the past
Make records err, and what we see doth lie
Made more or less as Thou art slow or fast.
This we do vow: our aim shall ever be
To chase Thee close in our pursuit of Thee.

Nathy O'Hora came across the poems while helping to sort
the observatory's archives in the 1980s and sent them to
Gemini, commenting that Perfect was 'as you may guess a
rather eccentric character. He was a great physicist but he
never forgot his education in classics and liked to read Latin
and Greek in the lunch hour.'[1]

come to an agreement with the Hailsham District Council that, if they
built a large number of houses in the village, council tenants could
occupy some of them at subsidised rents and observatory staff could
occupy the others at economic rents (in our case, just over two pounds
a week).'[Wa]

'As the clocks were all still at Abinger and Greenwich, I was an
embarrassment. Humphry Smith, the department head, and I were
the only Time Department staff and he had papers to write. I had
nothing. So he loaned me to the Electronics Department as an

Most boring card?

This postcard of the Time Service Control Room at Herstmonceux reached the final of a 'Most Boring Postcard' contest run by TV's Saturday morning show *Going Live*. Margaret Penston and Peter Andrews reported in *Gemini* that 'there was fierce competition for the title. The other finalists included an empty check-in area at Moscow airport, a round-about at Letchworth, Sutton Public Library, a fat man in a motel room, and the winner, an air drying cabinet. ... We were disappointed not to win the title (there is nothing less boring than a postcard which has won a most-boring competition). ... The final twist in this story is that we couldn't find a copy of the postcard anywhere – it has completely sold out!'[2]

unskilled wireman. We were setting up the Time Department in the West Building, which was only complete up to the top of the first storey, and the main entrances were unfinished. The only usable entrance was at basement level below the main entrance steps.'[Wa]

'On my first day, I was apprenticed to Arthur Cordwell, who took me to the stores. There, John Hobden issued me with a pencil, some paper and a pair of stout wellington boots. Arthur and I walked up to the West Building across the muddy fields, as no roads had been completed at that stage. Arriving at the West Building, Arthur pointed to what I thought was a flooded doormat well and said, "wash your wellies off in there before you go on to the woodblock floors." I stepped into what, in fact, was a drainage pit. The water came halfway up my thighs and I could see all the builders around looking at me uncomprehendingly. I splashed about a bit and said, "Come on in, Arthur, the water's lovely". I spent the morning soldering in an ill-fitting set of overalls while my trousers dried off in the clock cellars. I got funny looks from the builders for weeks afterwards.'Wa

'When we eventually got the Time Department finished and the clocks installed, we carried on receiving radio time signals from around the World and comparing them with our clocks. This, together with observations of star transits, enabled us to contribute to the deter-mination of Universal Time. It did mean that we needed to keep daily records of the clock performances, and of the measuring of Britain's main time signal transmitter at Rugby, call sign GBR. Though the rest of the observatory staff had the weekend off (except for the night observers), one Time Department staff member had to come in to receive GBR twice each day and to take clock measurements once each day.'Wa

Observations of the Moon also played an important part in the work of the Time Department, as Bill Nicholson explains:

'On the assumption that the Earth rotated once every twenty-four hours, time keeping was relatively simple, but when clocks became more accurate it became apparent that the Earth does not rotate uniformly, with the length of the day possibly varying by a microsecond or so, building up to several seconds over the years. To check on this, astronomers checked the position of the Moon. If the Earth slowed down, the Moon would not be in its predicted position. Errors in the Moon's position are shown up by recording the times at which a star's light is cut off as the Moon passes in front of it. The expected times of these "occultations" (which vary with position on the Earth) can be predicted very accurately, and any difference between the predicted and observed times is assumed to be caused by a change in the rotation of the Earth, in effect an error in the time system.'N

'Because of the labour and time involved, this was one of the areas in which amateur astronomers could play a vital part, and several hundred observers all over the world sent their observations to Herst-monceux to be combined together. But before they could do this they needed to know the predicted times at which to look out, and the Nautical Almanac Office had a large programme in sending out the predictions and processing the results that came back.'[N]

The same technique was done in reverse, when precise predictions of the Moon's position were used to find the exact position in the sky of astronomical objects which emit radio waves: 'An inverse use of occultations was in the determination, for the first time, of the position of a radio source on a photograph. At the time, in 1962, radio telescopes were not able to pinpoint radio sources very precisely, but could only point to a small area of the sky in which the radio source would be, but the small area (several minutes of arc in size) might contain many possible candidates and it was impossible to say which was the radio source.'[N]

'One particularly bright source, catalogued as 3C273, was timed by radio astronomers in Australia as it was occulted by the Moon on several occasions in 1962. Using the observed times, three positions of the Moon were plotted on a photograph of the sky. The edges of the Moon intersected to form, in navigational terms, a "cocked hat", and slap in the middle was the image which had to be, and in fact was, 3C273. This was the first such identification. One or two others followed, but fairly soon the radio telescopes had improved enormously

---

Watchmakers

Separate from the Time Department at Herstmonceux was the Chronometer Department, which in the 1950s included fifteen skilled watchmakers. The Department's job was to provide reliable timepieces – watches and chronometers – to the armed forces. In the Second World War the Department had tested and issued some 30,000 timepieces a year.

By the time of the Falklands War the Department had become the Chronometer Section of the Ministry of Defence, housed in the West Building at Herstmonceux. H E West reported to *Gemini:* 'The recent operations in the South Atlantic caused all three Services to make exceptionally heavy demands upon the Section. Subsequently, many of the staff worked very long hours for a period of five weeks. It is a matter of pride that all the demands upon the Section were met within the imposed time limits: the correct timepieces were issued to the vast range of members of all three Services who were connected with the Falklands Operation.'[3]

and the technique was no longer needed.'[N] 3C273 was subsequently identified by an American astronomer as the first 'quasar' to be discovered – an object that looks like a star in a photograph, but is actually a concentrated energy source as powerful as a million million ordinary stars.

'For measuring time, dependence on the rotation of the Earth has now been superseded by the use of atomic time as standard. With the realisation that time-keeping would come to be based on atomic time, the original plans for the West Building incorporated a large cellar, thirty feet deep, to accommodate the atomic clocks of that period, but even by the time the building was under construction atomic clocks had come down to suitcase size, and for many years the deep cellar lay unfinished and unused. As an aside, because of the importance of an accurate time system, the West Building housing the clocks was designed to withstand direct hits by any conventional bombs.'[N]

'The clocks for the time service, now of small suitcase size, were each housed in separate small rooms in the basement, and continually inter-compared with each other and also, by radio, with similar clocks elsewhere, in Paris, Washington, Tokyo, etc.'[N]

Compiling Tables

The Nautical Almanac Office, housed from 1958 in the new West Building, was an institution almost as famous as the observatory itself. *The Nautical Almanac* was (and is) a substantial volume of predictions of the positions of stars, Sun, Moon and planets throughout the year ahead, published annually since 1767 to enable people at sea to work out their position from observations they could make on board ship. Calculating these predictions, and those needed for a number of other publications, was an enormous task. At the time of the move to Herstmonceux the only computers available to tackle this mountain of mathematics were human ones, albeit with the aid of mechanical calculating machines. Electronic computers were yet to come. George Wilkins explains what attracted him to start work in the Office of which he would eventually become the head:

'Although I had chosen a book on astronomy as a school prize I had no intention of becoming an astronomer and my studies at Imperial College took me from physics to mathematics and into geomagnetism. At one time it seemed that my first job would be concerned with the design of electrical-power lines, but my PhD supervisor drew my attention to a vacancy in H M Nautical Almanac Office (NAO), which was already at Herstmonceux. He saw this as a possible route into the Magnetic & Meteorological Department of the Royal Greenwich Observatory, but for me the initial attraction was a job in computing that applied astronomy to the public service. The post also carried deferment from compulsory military service, but it was the location at the castle in proximity to the sea and the Downs that finally made the offer of a post irresistible.'[Wi]

'The Superintendent of the NAO was Donald Sadler (then known to the staff as DHS), and during my first few years in the NAO he gave me a series of jobs that gave me experience in the use of a variety of calculating machines and in the organisation of the work; I also had to learn about the techniques of editing and printing. The aims were to get results that could be trusted and to get them in an economical way; moreover, the printed numbers had to be right and the layout and style had to be such that the numbers could be used with the minimum of effort and risk of error. The production of the Almanacs was very much a team effort and Sadler made clear his displeasure with anyone whose work was not up to the expected standard. In all written memos the members of the staff were referred to by their initials; titles and given names were not used to show rank or status. Even the Astronomer Royal was known as the "AR".'[Wi]

'The NAO had a variety of desk calculating machines, the most popular of which was known as a Brunsviga; it was operated by turning a handle, but for complex tasks it was faster than electric machines. The bulk of the calculations were carried out on punched-card machines, each of which could perform a very limited set of operations. Their main virtue was that intermediate results could be passed from one stage to the next without anyone having to write them down and then reset them, as was the case with the desk machines. By the standards of the day they were very reliable, but a great deal of time had to be spent in applying checks to the results so as to pick up the errors made by the machines, their operators, and by those who designed the procedures and the "plugboards" that controlled the machines. At first all the results had to be keyboarded by the printer and so every member of the staff had to spend two hours each day reading proofs to find all the errors made by the printer or by our own staff in preparing the "copy". Each page was normally checked in some way by at least six different persons, but still the occasional error would slip through to the published volume. Eventually the NAO obtained an IBM card-controlled typewriter that produced copy that was good enough to be photographed to make the printing plate, but this could only be used economically if there were a large number of pages with the same layout.'[Wi]

'In 1957 it was arranged that I should spend a year in the USA to learn about the use of electronic computers, which were just becoming available commercially, and about celestial mechanics. Since I was to be away for only one year the Admiralty was only prepared to pay my foreign service allowance at the single-man's rate. Nevertheless my wife Betty, whom I had married in 1953, and our young son, Michael, went with me on the *Queen Elizabeth*. Just before I went to the USA, we had put in a bid for a DEUCE computer that was then being built by the English Electric Co. as an engineered version of the pilot model of the ACE (Automatic Computing Engine), which had been developed at the National Physical Laboratory. We were unsuccessful, and rumour had it that we were in direct competition with a bid for a new crane for the dockyard at Chatham!'[Wi]

'We were eventually allocated a Hollerith 1201 computer, which was considerably less powerful than the DEUCE and much more difficult to use than the IBM 650 computers which I had been using in the USA. It came without any software, and so I had to develop a set of routines that were equivalent to what is now known as an assembler. I even had to invent ways of representing and describing numbers based on 4 bits, and so in the early NAO Computer Circulars there are tables for "bi-octal" numbers (now known as hexadecimal bytes). We also had to develop all the basic routines for evaluating trigonometric functions etc. We did the initial development work on a computer in London, and I gave training lectures for other RGO programmers in the Chapel of the castle.'[Wi]

Dorothy Hobden was one of many who welcomed the arrival of the computer: 'Proofreading was a daily chore, which had to be done by every member of the NAO for about two hours a day, to check *The Nautical Almanac*, *The Astronomical Ephemeris*, *The Star Almanac*, etc. *The Nautical Almanac* was then produced on a special "Card Controlled Typewriter", before further checking and pasting up of special headers, etc. No one was sorry when it became possible to do all the production and checking by computer.'[H]

'The Hollerith 1201 was a valve computer and took up most of the ground floor. (I still have two of the old boards from it.) It had a memory of 1024 words — 64 rings of 16 words round a cylindrical drum. It was programmed in machine code and optimised by putting the next numbers required in the best place on the drum; for example to add one number to another, you put the second number two locations further round from the first.'[H]

'One evening I was working late on the 1201 when everything ground to a halt. I couldn't find what the problem was, so shut down, and went to watch the evening cricket match. Next day I discovered the cause: simply that the line-printer had run out of paper. I didn't let that happen again!'[H]

'It soon became clear that the 1201 computer was not capable of doing many of the NAO data-processing jobs, and it was even less suited to research projects. I acquired a copy of a program from the Jet Propulsion Laboratory for the evaluation of the coordinates of the Moon and used it to teach myself Fortran so that I could run the program on an IBM 7090 computer in London. We soon began to run other jobs in this way and to press for a replacement for the 1201.'[Wi] 'From 1962-64 I used the IBM 7090 up in Newman St, London. It still used punched cards and paper tape, plus magnetic tapes which were a new advance.'[H]

In the 1960s, ordering a new computer was a major exercise:

'I spent a large part of one year in a computer specification and evaluation exercise involving the Post Office Technical Support Unit and Admiralty O & M. At the end of it we went out to tender to four companies. I put great weight on software and on compatibility with the US Naval Observatory and other astronomers, and so I favoured an IBM 360, but the computer selected by the team was an ICT 1909, which undoubtedly had more hardware power for its price. We insisted on an acceptance trial that lasted a week and during which the computer had to demonstrate its reliability and its ability to run four programs together. We subsequently expanded the system by replacing the central processor and adding disc drives discarded by the Atlas Computer Laboratory, but although the system worked well we were never able to share software easily with other groups.'[Wi]

'The ICT (later ICL) computer arrived in the mid sixties. I

remember a painful moment after working late on it one evening. I
went to the clubhouse (in the dark) and made for the light on the
front door, only to fly headlong over a pole put across a gateway by
the farmer, to keep the sheep out. I spent an uncomfortable night,
and next morning at the dentist. But I did get compensation through
the Union, with which I bought a kettle and an iron. Eventually we
didn't have to "buy British" and changed to DEC VAX computers.'[H]

Satellites and Lasers

'While I was in the USA in 1957-58 the first artificial satellites were
launched, while at Herstmonceux the West Building was completed. When I arrived back I was pleased to find the NAO in its new
offices, but I was very disappointed to find that Woolley had insisted
that the NAO should abandon the satellite prediction service that it
had established; the responsibility was transferred to the Royal Aircraft
Establishment. During 1970 the Astronomy Committee of the Science
Research Council decided to stop the publication of our principal
publication, then known as *The Astronomical Ephemeris*, but I was allowed
to speak at the next meeting and I was able to get the decision
rescinded.[Wi]

'In the early 1970s, the NAO was party to a university proposal
for a UK lunar laser ranging system to be sited at Sutherland in South
Africa, but the proposal was rejected when it was found that all the
costs would have to be met by the UK. Some years later the RGO
supported a revised proposal for a satellite laser ranging system, and
we eventually found ourselves taking the lead role.'[Wi]

The lunar ranging system would have involved beaming pulses of
laser light from the Earth onto reflectors that had been left on the
Moon by astronauts, and timing how long it took for the reflected
pulses to return to Earth. The resulting information makes it possible
to calculate an extremely precise value for the distance between the
Earth station and the lunar reflector. The satellite laser ranging system
(SLR) that eventually came into service at Herstmonceux in 1983
works the same way but on a smaller scale. Laser pulses are fired at
satellites passing over Herstmonceux. By timing the return of the
reflected pulse from the satellite, the distance from the station to the
satellite can be found to an accuracy of a centimetre or two.

The SLR was run by the observatory's Time Department and was the only part of the RGO to remain at Herstmonceux when the rest of the observatory moved to Cambridge. At the time of writing (1999) it is still in operation day and night – provided the sky is clear and there are no aircraft in the line of fire. The SLR is part of a global network of such stations whose results provide information about, amongst other things, continental drift – the slow movement on the Earth's surface of one continent relative to another.

Two of the observatory's earlier 'long-term projects (occultations and mapping by overlapping photographic plates) have now been overtaken by astronomy from satellite observatories. With reflectors placed on the Moon by astronauts, observation of these, and other satellites, by the Satellite Laser Ranger telescope has superseded the dependence on occultations. And the laborious photographing and measuring of the plates, which occupied some twenty years, has recently been carried out in some eighteen months by the HIPPARCOS satellite.'[N]

References

1   *Gemini* 13, February 1985, p11

2   *Gemini* 27, March 1990, p25

3   *Gemini* 2, July 1982, p5

5

Moving on

In the long history of the Royal Observatory, the Sussex years form only a small part. The golden age (if such it was) that started in the 1950s began to fade long before the observatory at Herstmonceux finally closed its doors in 1990. That closure, less than forty years after the observatory fully opened, and the removal of the RGO to Cambridge brought disruption, acrimony, sadness and bitterness. But it was not unexpected. As George Wilkins explains, changes in the observatory's role meant that the writing had been on the wall for many years:

'The year 1965 saw the transfer of responsibility for the RGO from the Admiralty to the newly-formed Science Research Council (SRC). The consequential increase in funding for astronomy led to a welcome increase in the complement of the RGO, but to an even greater increase in the amount of time and paper that was involved in non-scientific activities such as the annual cycle of estimates and reports. The more significant change was that the RGO was no longer regarded as an observatory whose main functions were to carry out astronomical programmes of observation and to provide national and international services for navigation and time. Instead the RGO's first priority became the support of research in universities; the conduct of its own projects came second, and services soon became a poor third. Woolley managed, however, to continue to develop the research programme in association with the Institute of Astronomy at the new University of Sussex. His successors were less successful and the RGO's own programme was gradually reduced and subjected to frequent external review.'[Wi]

'I suspect that few of us who watched the building of the dome for the Isaac Newton Telescope in the mid-1960s realized that the future of the RGO would be so closely linked with that telescope's success or failure. Unfortunately it came ten years too late, and the RGO did not, at first, give sufficient priority to its development and operation. As a consequence, the task of building and operating the

Hemisphere Observatory (NHO) was almost denied to the RGO.'[Wi]

'The decade of the seventies saw major changes in the role and administration of the RGO. Woolley retired at the end of 1971 and the title of Astronomer Royal was conferred on Martin Ryle, a radio astronomer at the University of Cambridge, rather than on Margaret Burbidge, who was belatedly appointed as the next Director of RGO. Our IPCS Branch Secretary, Joy Penny, claimed that this was yet another indication of SRC's intention to take away the national status of the observatory and, unfortunately for the RGO, she has been proved right.'[Wi]

'The 1975 Tercentenary celebrations can now be seen to have occurred at the time of a crucial change in the role of the RGO. The decision to move the Isaac Newton Telescope (INT) to La Palma was made for political reasons, rather than on technical or financial grounds, but I did not realize at the time that it would eventually lead to the complete abandonment of Herstmonceux as the site of the RGO. It was ironic that the dismantling of the INT took place in the same year (1979) as a new wing of the West Building was brought into use for the staff who were engaged in astrophysical research.'[Wi]

'Soon after Alec Boksenberg became Director in 1981, the RGO suffered a major redundancy exercise that reduced the complement by about 25 per cent. I am thankful that on his first day in office he relieved me of the additional duties as Deputy Director that I had been given by Graham Smith during his last year as Director. (Instead he appointed a full-time scientific administrator and so was able to spend much of his time away from Herstmonceux.) It was a wretched task to identify the staff who should receive a "brown envelope" containing an offer of "voluntary premature retirement", especially as my Division did not have the priority work associated with the new observatory and telescopes on La Palma. The Greenwich Time Service suffered particularly badly and we lost experienced staff with specialist knowledge from the NAO.'[Wi]

'The brown-envelope exercise was followed by a series of reviews, culminating in the decision by SERC (the Science and Engineering Research Council, successor to the SRC) to move the RGO from Herstmonceux to another site. Almost all the staff, apart from a few senior people, were strongly opposed to this decision and some of us wrote to our local MPs to try to gain their support. They appeared sympathetic, but it appears that they forwarded our letters to the SRC,

---

Who needs old plates anyway?

Every photograph taken using any of the instruments at Herstmonceux was preserved in the RGO's plate archive. As the time approached to move the archive to Cambridge, Chris Thoburn gathered some facts and figures and asked 'What weighs 18 tonnes and will occupy some 93 m³ in the new RGO building at Cambridge? Answer: the RGO plate archive.'

'The plate archive contains plates exposed as long ago as 1888 ... at Greenwich ... and as recently as 1988 on the 26-inch refractor at Herstmonceux. The collection includes some 58,000 direct plates from the 26-inch Thompson Refractor (Greenwich and Herstmonceux, 1896-1988), 21,000 13-inch astrographic plates (Greenwich and Herstmonceux, 1888-1987), 22,000 solar plates (Greenwich and Herstmonceux, 1918-1979) ...'

'How do you move 120,000 glass plates? You need at least 1200 specially designed boxes, several trucks, a spare six months for the packing and a strong back.'[1]

---

new Northern with the result that we were admonished for our actions. Two ladies who had written on behalf of the trade unions were even subjected to formal disciplinary proceedings. The proposal was widely criticised in the press and the SERC was unable to obtain the approval of the astronomical community for the move, but it went ahead and decided to move the RGO from its fitting home at Herstmonceux Castle to the garden of a small Victorian observatory in Cambridge!'[Wi]

'I was fortunate in that I was due to retire before the move would actually take place, but many of the staff faced a major disruption to their lives, and some decided not to leave Sussex, rather than uproot their families. There were many regrets when we had our last Xmas Lunch in the castle and when, a few months later, we held a "Farewell to the Castle" party just before it was handed over to the developer who had bought the estate. I was able to retain the use of my office in the West Building until the move to Cambridge took place in April 1990. I cleared my desk on April 4 and in the evening played in the RGO table-tennis team in a drawn league match, which was held in the Clubhouse. My diary records that "I won two sets!!" – not bad for a veteran!'[Wi]

Postscript

The move to Cambridge in 1990 turned out to be merely a stay of execution. Seven years later the future of the Royal Greenwich Observatory again became precarious with the announcement that its funding body, now called the Particle Physics and Astronomy Research Council, was withdrawing its support. Key functions of the observatory would transfer elsewhere and it would no longer retain its Cambridge building.

When the Cambridge site closed in the autumn of 1998, history came full circle. The institution's one remaining function, its 'Public Understanding of Science' service, was moved back to Greenwich where the observatory had begun life 323 years before. The title of the historic establishment there reverted from 'Old Royal Observatory' to its original 'Royal Observatory Greenwich'.

At Herstmonceux the future also looked bleak in 1990, particularly when the development company that had purchased the castle and estate fell into liquidation. But by 1994 the outlook was much brighter. Castle and estate had been acquired by a benefactor for use by Queen's University, Ontario, Canada as an International Study Centre, and the first students were in residence. The Equatorial Group was made available to Science Projects Ltd, a charitable company experienced in presenting science to the public, and soon came back to life as Herstmonceux Science Centre. Today, the Centre's many visitors enjoy the opportunity not only to experience 'hands-on science' for themselves, but also to savour the unique atmosphere of what was once one of the world's leading astronomical centres.

References

1   *Gemini* 27, March 1990, p26

Index